Thoreau's Botany

Under the Sign of Nature: Explorations in Environmental Humanities
Serenella Iovino, Kate Rigby, and John Tallmadge, Editors
Michael P. Branch and SueEllen Campbell, Senior Advisory Editors

Thoreau's Botany

THINKING AND WRITING WITH PLANTS

James Perrin Warren

UNIVERSITY OF VIRGINIA PRESS
CHARLOTTESVILLE AND LONDON

University of Virginia Press
© 2023 by the Rector and Visitors of the University of Virginia
All rights reserved
Printed in the United States of America on acid-free paper

First published 2023

9 8 7 6 5 4 3 2 1

Library of Congress Cataloging-in-Publication Data

Names: Warren, James Perrin, author.
Title: Thoreau's botany : thinking and writing with plants / James Perrin Warren.
Description: Charlottesville : University of Virginia Press, 2023. | Series: Under the sign of
 nature : explorations in environmental humanities | Includes bibliographical references
 and index.
Identifiers: LCCN 2023003604 (print) | LCCN 2023003605 (ebook) | ISBN 9780813949475
 (hardcover) | ISBN 9780813949482 (paperback) | ISBN 9780813949499 (ebook)
Subjects: LCSH: Thoreau, Henry David, 1817–1862—Knowledge and learning. | Thoreau,
 Henry David, 1817–1862—Criticism and interpretation. | Botany in literature. | Plants
 (Philosophy) | Ecoliterature, American—History and criticism. | Botany—United
 States—History—19th century.
Classification: LCC PS3057.N3 W37 2023 (print) | LCC PS3057.N3 (ebook) |
 DDC 818/.309—dc23/eng/20230411
LC record available at https://lccn.loc.gov/2023003604
LC ebook record available at https://lccn.loc.gov/2023003605

Cover art: Background, *Thoreau's Journal,* Charles W. Jefferys, 1896 (courtesy of Anthony
Allen); book, Julia Tsokur/shutterstock.com; leaf and acorn design, BTSK/shutterstock.com

To Julianne L. Warren

One day in a row

Contents

Preface

In February 2018, I visited my friend Barry Lopez at his home in Finn Rock, Oregon. The visit came about because I was reading drafts of Barry's big book, *Horizon,* a work of nonfiction that he had been researching and writing for nearly twenty-five years. It also came about because I knew that Barry was approaching the end of a seven-year journey with prostate cancer. As it turned out, this would be my last chance to sit with him and talk—about his work, about wild animals, about travel into remote places, and about his love for the McKenzie River and the Douglas fir forests he had made his home for fifty years.

For three days, we sat together, talking, sometimes with recording equipment running, sometimes off the record. Barry was in deep pain, physical pain. Sometimes he called it "knee-buckling." But he insisted that we sit and talk, and in those three days he never complained about his life being cut short, the projects he envisioned for the future coming to a sudden halt. Mostly he was worried about *Horizon,* that the book wouldn't live up to what he wanted it to be. He could darken in talking about it, and then he would brighten when we talked about a projected road trip from Key West to Prudhoe Bay, about roads and vehicles, or about encounters with wolverines.

The visit taught me a lesson about how a dedicated writer might face the certain end of writing. In Barry's case, he kept going. He continued planning new projects, learning new material, working his prose, honing his sentences, experimenting with rhythms and images. He kept his eye on the river and on the woods. With the constant support of his wife, Debra Gwartney, he kept going, placing his faith in his many writing practices.

Insights from that visit have stayed with me the last four years, during the main work of researching and writing *Thoreau's Botany.* Thoreau's final decade is fascinating and admirable, not least for the faith he placed in the work he found himself doing. It could not have been obvious where the botanical studies of the 1850s were going. They would enter his writings, to be sure, as *Thoreau's Botany* presents in detail. But for what purpose, with what actual goal? As it turned out, they became all absorbing, the main work of every day, the desire taking him into the woods and fields and swamps and streams within a ten-mile radius of Concord, Massachusetts. As the excursions proliferated, they grew in significance. His idea was to make wholes out of the parts, to see the ways in which the world was a whole and how it was often being taken apart. Plants became the signifying center of his days.

Thoreau was hard at work on several related botanical projects when, in December 1860, he went out into the woods in cold, wet weather and caught a cold, which descended into his chest and ultimately led to his death, at age forty-four, in May 1862. As his sister Sophia could attest, he worked especially hard at preparing essays for publication in the final months, and he was working on the last section of *The Maine Woods* to the very end. When he was no longer able to make his botanical excursions, friends brought flowers and fruits to his sickbed.[1]

Thoreau's last years have been the subject of debate for decades, and in the past thirty years scholars and critics have begun to appreciate the posthumous publications, unfinished manuscripts, and Journal entries that occupied the writer after *Walden* (1854). One problem with the works of the final decade is that few readers have seen how Thoreau's plant studies impact his thinking and writing.[2] Instead, we have general literary appreciations for the botanical works in the name of Thoreau's attention to material reality and his focus on the concrete details of nature. The botanical works are much deeper than these basic accounts, and they deserve to be analyzed and discussed fully. Thoreau was a first-rate field botanist and a prescient student of forest ecology and evolutionary theory. At the same time, he was Thoreau—a critical, intelligent, faithful thinker and literary writer.

In the introductory chapter, "Thoreau's Botanical Turn," I chart how Thoreau's dedication to botany takes on a dramatic literary life. The pages of Thoreau's Journal are full of records, especially of the writer's excursions around Concord. In 1850, the Journal changes from a literary workshop for drafting material to an independent writing project. At the same time, during the winter of 1850, Thoreau begins the disciplined, focused study of botany that he develops through 1851. He had worked on plants earlier, but in 1850–51 he begins studying botany systematically. He studies the botany manuals of Jacob Bigelow and Asa Gray, along with the works of older botanists like Carl Linnaeus, André Michaux, and François André Michaux. Beginning in

1850 and continuing for over a decade, the Journal entries record the plants observed, described, collected, and identified on the nearly daily walks that Thoreau made. In addition, the entries are records of ways of knowing, reflections on science and poetry, and reflections on his own practices as a writer, a "walker," and a "traveler." They are faithful accounts of his emotional life and of his doubts concerning the meaning of his life; he interrogates himself constantly, asking himself hard questions about the state of the world and his vocation as a writer in the world. He finds faith near the ground and in the plants, in their recurring processes of growth and reproduction.

In part 1, "Reconstructing the Botanical Excursions," three chapters show the role of botany in Thoreau's daily walks and extended travels. Already in his first published essays and in *A Week on the Concord and Merrimac Rivers*, Thoreau offers travel narratives that he called "excursions." An excursion narrates a journey, usually out to a previously unknown place and then back home. This creates a dynamic of crossing a boundary into the new and unknown, then returning home to the familiar or newly unfamiliar place of origin. The dynamic pattern of the excursion provides a window into Thoreau's writings, since he is preeminently a traveler, even in Concord. He is a writer who travels.

In published works, the search for a faithful record that we see in the 1850–51 Journal leads toward the botanical excursion as a narrative form that brings culture and nature into productive, dynamic relationships. This type of narrative engages the first-person narrator/character, the place being experienced and explored, and the ways of knowing the place. There are limits to these writerly elements, and Thoreau is constantly experimenting with them to expand his ways of knowing and representing the world. He is always asking how he can be a faithful traveler and writer, how he can bring plants and humans into meaningful relationship. The three chapters of part 1 focus on the published works *The Maine Woods*, *Cape Cod*, and *Walden*, arguing that the botanical excursion becomes a defining form of narrative in the mature works of the 1850s. This form develops through successive, overlapping, multiple excursions; multiple versions of a text; and multiple ways of knowing and experiencing a place.

In part 2, "The Broken Task & the Faithful Record," the late natural history projects appear in recently discovered and edited form. The first of three chapters focuses on Thoreau's Kalendar project as a way of reading the Journal, showing how the Kalendar and Journal function together as faithful records of Thoreau's botanical excursions in the immediate region of Concord from 1852 to his death. The entries in the Journal connect Thoreau and Concord to the seasonal world and make the world of Concord into a more-than-local place. The Journal creates openings for Thoreau, giving him new possibilities for writing and recording his experiences.

The second chapter discusses the manuscript edition of *The Dispersion of Seeds* and the writing process as practices of succession and dispersion. The ecological process of succession provides the means for developing themes of spatial dispersion and temporal succession. Together, dispersion and succession explain how landscapes change, how one species of forest tree replaces another. Patterns of succession are in turn affected by human agency, by the landowners and their practices of cutting, burning, planting crops, and allowing the forests to regrow. Thoreau's faith in his own work lies in *phusis,* the in-dwelling dynamism of the earth, because he sees his writing as part of the dynamic processes all around him.

The third chapter suggests that the edited volume *Wild Fruits* is a dynamic, performative hybrid of many fragments, an experimental form of writing based on repeated excursions and observations of plants. It gives the most developed example of how Thoreau's botanical work of the last decade delivers new thinking and writing, and it is the best indication we have of how his seasonal thinking yields a written set of botanical excursions and reflections. Thoreau's botany proposes a more intimate relationship between human beings and the land they occupy, manage, and cultivate. *Wild Fruits* is persistently unsettling, merging wild and cultivated fruits to develop fresh ideas of growth and transformation.

In the epilogue, I reflect on the importance of Thoreau's botany and his botanical excursions for readers in the Anthropocene, especially during the two full years of the coronavirus pandemic. The epilogue is meant to suggest directions that our reading of Thoreau's botany might take in the present and immediate future.

Even though Emerson characterized Thoreau's final years as "a broken task," *Thoreau's Botany* describes the faithful practices of a mature, original writer and thinker. In that regard, Thoreau reminds me strongly of Barry Lopez, working resolutely and brilliantly under the certainty of pain and death. Both are writers who travel; both focus on ethics and landscapes; both seek to understand all the inhabitants of a landscape as spirited, living neighbors and teachers; both face a sudden catastrophic end with faith and conviction. Both keep going, even now.

Acknowledgments

This book began in the woods of western Virginia. My good friend and colleague John Knox, an expert teacher of field botany, allowed me to audit a six-week spring course with him in 2005. For five more years after that, he and I cotaught a spring course, Field Biogeography: Science and Literature. We spent our days identifying hundreds of flowering plants and trees in the field, discussing the discipline of biogeography, and reading works by Charles Darwin, Alfred Russel Wallace, Barry Lopez, David Quammen, and Robin Kimmerer. Along the way, we became coteachers in a deep sense. Since that time, John has kept me going with our friends Kirk Follo, Larry Hurd, and Helen I'Anson, colleagues at Washington and Lee University and intrepid walkers.

The students in those spring courses were a constant, refreshing inspiration. We also welcomed fellow walkers in Robin Kimmerer and Barry Lopez. To have Robin teach us a three-hour lab on mosses was unforgettable. To have Barry walk with us for two weeks, remarking on the hardused landscapes of Virginia, was a way of seeing our world with new eyes.

Friends from the Association for the Study of Literature and Environment (ASLE) were supporters long before botany became my primary focus. John Tallmadge, John Lane, and Michael Branch have maintained a steady interest in my work for decades, an interest that has continued to this book as well. ASLE has been an intellectual home for me since 1997; without it, this work would have never been conceived, much less brought to fruition. Through ASLE, I've been blessed with friends like Kurt Caswell, Rochelle Johnson, Lauren LaFauci, Mark Long, Ian Marshall, Annie Merrill, Jake McGinnis, Lance Newman, and David Taylor. Many more ASLE friends deserve my gratitude, too, but I especially want to single out Amy McIntyre, who has shepherded the organization expertly for years.

This book owes deep debts to Thoreau scholars, past and present, most of whom have no idea how much they have meant to my studies. I have been grateful for encouraging words and editorial help from Jeff Cramer, Kathleen Kelly, John Kucich, Alex Moskowitz, and François Specq. At the 2018 Thoreau conference in Gothenburg, Sweden, organized by the unflappable Henrik Ottenberg, I was inspired by generations working on Thoreau, from elders like Robert Sattelmeyer and Joseph Moldenhauer to younger scholars like Rochelle Johnson and Kristen Case. Then there are scores of Thoreau critics, many of whom I've never met, whose work has led me expertly: Ray Angelo, Branka Arsic, Lawrence Buell, Cristen Ellis, William Howarth, Lance Newman, Daniel Peck, Sandra Petrulionis, David Richardson, David Robinson, William Rossi, Robert Thorson, Elizabeth Witherell, to name the most prominent. Without the editorial genius of Bradley Dean, the task I set myself here would be broken. Thoreau scholars, named and unnamed, can be proud of the community they have forged.

The most direct inspiration for this study is Laura Dassow Walls. Her early books on science in Thoreau and Emerson brought interdisciplinary scholarship on science and literature to vivid life. Then she brought us Alexander von Humboldt, and, after that, Thoreau, in the definitive biography and in a host of precisely and generously reasoned articles. Always, Laura has shown and shone a light on Thoreau, in his own time and in ours.

I thank the librarians at Washington and Lee University, especially Elizabeth Teaff, for loaning me the scholarly edition of Thoreau's Journal—a godsend, truly.

I am grateful to Angie Hogan, Acquiring Editor, to Ellen Satrom, Managing Editor, and to the other personnel at the University of Virginia Press. To the two anonymous referees, many thanks for excellent advice in improving the manuscript. For expert copyediting, my thanks to Emily Shelton.

Finally, and always, I am most grateful to my partner, Julianne, whose example leads me forward, every day and for many days in a row.

Abbreviations

CC   Henry David Thoreau. *Cape Cod*. Edited by Joseph J. Moldenhauer. Princeton, NJ: Princeton University Press, 1988.

Corr   Henry David Thoreau. *The Correspondence*. Edited by Robert N. Hudspeth, Elizabeth Hall Witherell, and Lihong Xie. *The Writings of Henry D. Thoreau*. 2 vols. Princeton, NJ: Princeton University Press, 2013–.

EEM   Henry David Thoreau. *Early Essays and Miscellanies*. Edited by Joseph J. Moldenhauer and Edwin Moser, with Alexander Kern. Princeton, NJ: Princeton University Press, 1975.

Exc   Henry David Thoreau. *Excursions*. Edited by Joseph J. Moldenhauer. Princeton, NJ: Princeton University Press, 2007.

Faith   Henry David Thoreau. *Faith in a Seed: The Dispersion of Seeds, and Other Late Natural History Writings*. Edited by Bradley P. Dean. Washington, DC: Island, 1993.

J   Henry David Thoreau. *The Journal of Henry David Thoreau*. 14 vols. Edited by Bradford Torrey and Francis H. Allen. New York: Dover, [1906] 1962. (Volumes indicated by roman numerals.)

MW   Henry David Thoreau. *The Maine Woods*. Edited by Joseph J. Moldenhauer. Princeton, NJ: Princeton University Press, 1972.

PJ   Henry David Thoreau. *Journal. The Writings of Henry D. Thoreau*. Edited by Elizabeth Hall Witherell, Robert Sattlemeyer, and Thomas Blanding. 8 vols. Princeton, NJ: Princeton University Press, 1981–. (Volumes indicated by Arabic numerals.)

RP   Henry David Thoreau. *Reform Papers*. Edited by Wendell Glick. Princeton, NJ: Princeton University Press, 1973.

W   Henry David Thoreau. *Walden*. Edited by J. Lyndon Shanley. Princeton, NJ: Princeton University Press, 1971.

WF   Henry David Thoreau. *Wild Fruits: Thoreau's Rediscovered Last Manu-script.* Edited by Bradley P. Dean. New York: W. W. Norton, 2000.

Wk   Henry David Thoreau. *A Week on the Concord and Merrimack Rivers.* Edited by Carl F. Hovde, William L. Howarth, and Elizabeth Hall Witherell. Princeton, NJ: Princeton University Press, 1980.

Thoreau's Botany

Thoreau's Botanical Turn

CRITICAL PLANT STUDIES AND PLANT-THINKING

Over the past thirty years, students of critical animal studies have developed new ways of discussing the relationships between the human and more-than-human world. More recently, the interdisciplinary wave has emerged in critical plant studies. The philosopher Michael Marder's *Plant-Thinking: A Philosophy of Vegetal Life* (2013) stands as a seminal text in the field. Among many writing projects, Marder is editing a new series of interdisciplinary publications called *Critical Plant Studies.* In addition to philosophical and literary studies, such recent popular works as Richard Powers's novel *The Overstory* (2018) and David George Haskell's *The Song of Trees* (2017) attest to the power of new approaches to plants. Perhaps the best example is Suzanne Simard's *Finding the Mother Tree: Discovering the Wisdom of the Forest* (2021), a book that combines personal narrative, three decades of scientific research, and startling insights into the subtle networks that form forests. These works bring the question of our relationship to plants to a new tipping point.[1]

In *Plant-Thinking*, Marder defines his principal concept of "plant-thinking" in four ways: 1) "the non-cognitive, non-ideational, and non-imagistic mode of thinking *proper to* plants" (10); 2) human thinking *about* plants; 3) human thinking as altered by its encounter with the vegetal world, becoming less human and more plant-like; and 4) "the ongoing symbiotic relation between this transfigured thinking and the existence of plants" (10). In the combination of these four senses, "plant-thinking" is above all an encounter with the vegetal other. As a sequence, the four definitions create an implicit narrative of contact, intersection, and transformation. By taking account of the mode of thinking proper to plants and then by altering our thinking to bring it into accord with that vegetal mode of thinking, we humans can create the ongoing symbiotic relationship with plants that will transfigure our lives and our futures on the planet.

There is ample evidence to support the proposition that Henry David Thoreau engaged in all four modes of plant-thinking in his writing, especially in the work he produced after he turned to the rigorous study of plants in 1850–51. Thoreau's botanical turn was decisive, and it continually deepened his thinking and writing until his death in 1862. This period of some twelve years coincides with the beginning of his organized collecting of specimens for an herbarium and with his increasingly detailed and knowledgeable observations about flowering plants and trees in Concord and beyond. It matches the timing of his deep reflections on the relationship between science and poetry as ways of knowing and as modes of expression. It overlaps with the last revisions of *Walden*, the drafting of *Cape Cod*, and the incomplete drafting of *The Maine Woods*. It underpins the works we think of most capaciously as "the Kalendar project," including Bradley Dean's editions of *The Dispersion of Seeds* and *Wild Fruits*. Finally, it may add telling details to enrich our interpretive sense of Thoreau's empirical holism and his growing interest and respect for Native American ways of knowing.

Marder's argument in *Plant-Thinking* is in two parts. Part 1, "Vegetal Anti-Metaphysics," is a posthumanist deconstruction of the Western philosophy of the plant. The deconstruction also takes place in two parts. "The Soul of the Plant" recounts the history of Western metaphysics as a resolute misreading of vegetal life as an "other" and therefore lesser state of existence. For example, Aristotle sees the plant as "incomplete" because it is not goal directed, while Marder reads "incompleteness" as meaning "open-ended," so the quality becomes a positive attribute (23–24). In a later phase, Hegel privileges the "interiority" of the animal and human soul, so that the fundamental exteriority of the plant leads to its spiritual instrumentality, always understood as subservient to the needs and uses of human beings (24–30).

In reading Western metaphysics, Marder argues that the plant—or "vegetal being"—is a synecdoche for nature itself, *phusis*, which is defined by a fundamental and elusive vitality. Although the history of Western philosophy renders vegetal life as inert and mute matter, Marder suggests that the contemporary philosophical defense of plants "bears upon all of *phusis*, without running the risk of replicating the abstract, general, and indifferent metaphysical thinking enamored with totalities, such as nature or indeed the environment" (31). Marder specifically revises the ancient concept of *phusis* to define nature as dynamic, as marked by change and becoming. This idea of vegetal being as an embodiment and figure for *phusis,* the fundamental vitality of nature, will be of signal importance to understanding Thoreau's plant-thinking. Indeed, Thoreau's plant-thinking develops over the course of the last decade of his life in many dynamic, powerful, and elusive ways. In that sense, Thoreau's botany is a forerunner for the new approaches to the human relationship with plants we see in our present time.

For Marder, the deconstruction of Western metaphysics is also reconstructive and instructive. For instance, he reads "vegetal desire" as nutritive, suggesting a provisional status for Aristotle's concept of the vegetal or nutritive soul, which Marder then uses to argue for specific aspects of commonality with animals and humans in the processes of nourishment, growth, and procreation (48). In the second part of Marder's deconstruction, "The Body of the Plant," he argues that the plant silently undercuts the idealism of Western philosophy, enacting the deidealizing of human thought and existence that we associate with materialism (60). Nietzsche and Heidegger come strongly into play in this section of the argument, for the Nietzschean transvaluation of values leads Marder into a nonanthropocentric phenomenology that explicitly critiques the idealizations of Heidegger's *Dasein*. Throughout part 1, Derrida is the clear inspiration for Marder's strategies of reading and argumentation.

In part 2, Marder develops three aspects of vegetal existence: time, freedom, and wisdom (or intelligence). In the chapter on temporality, Marder discusses seasonal changes, growth, and cycles of repetition and reproduction. In the (very dense) chapter on freedom, he undermines the truism of the plant's fixedness, suggesting ultimately that the plant's freedom from need, the mark of vegetal "indifference," grounds a playful exuberance that is most akin to the "material freedom of imagination" (146). Despite such hopeful conclusions, this is a difficult argument to make, and Marder admits that "plant liberation" requires persistent "plant-thinking piercing through layer after layer of the idealist repression weighing upon it" (148). Finally, under the theme of wisdom, Marder gives versions of unconscious memory, intentionality of nourishment and reproduction, and a form of thinking ("it thinks" rather than "I think") that embraces life processes more broadly conceived than consciousness; the works of Henri Bergson, Gregory Bateson, and Gilles Deleuze underpin these arguments. The upshot is that Marder sees plants as fitting within their surroundings in ways that conjoin the plant and its other in a harmonious unit of survival, rather than an antagonistic relation of self-conscious subject and distant object.

In a brief epilogue, Marder proposes ten ethical "offshoots" from his arguments. The most important is the first: that "plant-thinking is plant-doing. . . . All plant-thinking actively takes the side of the plant and works for the sake of the plant" (181). Ethics, Marder contends, is an offshoot of vegetal life and of plant-thinking; vegetal life deserves respect from human beings in all our relationships with it. In becoming more consonant with plant-thinking, our ethics will become more attuned to the ways in which plants form communities and engage with organic and inorganic others. Such an ethics creates a place for an open-ended temporality, a continuing learning from plants that does not drive toward final conclusions. Once again, we can imagine that both this ethical aspect and the open-ended temporality it creates will become essential in a reading of Thoreau's botany.

The nagging problem with plant-thinking, as with much Derridean philosophy, is language. Marder treats "the language of plants" in chapter 2 (74–90), but his account of how vegetal life "expresses itself without resorting to vocalization" (75) seems to sidestep the problems of representation and the difference between human language and nature. If, as Marder suggests, plants express themselves materially, how would human language become more consonant with this aspect of vegetal life? The alienations of thought and language are not a convincing answer to that question, but they tend to fill the pages of *Plant-Thinking.*[2]

Marder finds a more positive answer in the leaf as "the very embodiment of supplementarity" (81) and the seed as a "singular plural" of Derridean dissemination (89). But these paradoxical images only hint at Marder's embrace of a decentered, disseminating theory of language. Later, in the chapter on the time of plants, Marder returns to the leaf and seed and to the subject of expression. The result, however, is a deferral, a figuration of "the plants' proto-writing" (116). When Marder considers Heidegger's humanist, metaphysical critique of plants and animals as lacking a "world" because they lack language, he asks whether the spatial forms of plants may constitute their "free opening unto being right within their environments" (129). Unfortunately, this question, as telling as it is, does not solve the problem of language and its limitations. If vegetal life suffers because of the insufficiencies of our language, how does its essential voicelessness contribute to that suffering? Indeed, the eighth offshoot in Marder's epilogue is that "the plant's absolute silence puts it in the position of the subaltern" (186). Marder realizes that there are difficult hurdles for those who seek to speak for plants without objectifying them, but in this brief prolegomenon he does not develop a method for overcoming these fundamental obstacles. It seems that the best way forward is to try to think and act like a plant, which seems to mean thinking and acting like a posthumanist, postmetaphysical philosopher.

The difficulties faced by Marder in outlining the elements of plant-thinking in terms of Western metaphysics may well be insurmountable, insofar as philosophical or linguistic thought is concerned. It may well be, however, that interdisciplinary thinking can approach the elements of plant-thinking without falling into the problematic prison-house of language. This possibility looms large in Thoreau's writing, especially in the botanical excursions of the Journal. Rather than stop with the admittedly difficult obstacles in developing a philosophy of vegetal being, we should take Michael Marder's work as a useful, generative outline for reading and understanding Thoreau's botany.

PLANT-THINKING IN THOREAU'S 1851 JOURNAL

Thoreau's botanical turn begins in the summer of 1850, but it is a process that requires several years to bear significant fruit. In a well-known Journal entry of 4 December 1856, Thoreau looks back on his self-education as a practicing botanist. He had begun studying local plants as an undergraduate at Harvard University, using Jacob Bigelow's popular manual *Florula Bostoniensis* (1824, 1840). In retrospect, Thoreau sees clearly how far he had to go. He began by "looking chiefly for the popular names and the short references to the localities of plants, even without any regard to the plant." He learned names, "but without using any system, and forgot them soon." He resisted the urge to collect specimens, and he paid little attention to plants in his family home. "But from year to year we look at Nature with new eyes." By his account, about 1850, he began "attending to plants with more method, looking out the name of each one and remembering it," collecting them and bringing them home in his straw hat, "which I called my botany-box." And he fought against the frustration of the novice: "I remember gazing with interest at the swamps about those days and wondering if I could ever attain to such familiarity with plants that I should know the species of every twig and leaf in them, that I should be acquainted with every plant." Reflecting further on the growth of his knowledge, he rejects systematic study ("the most natural system is still so artificial") and instead wants to "know my neighbors, if possible,—to get a little nearer to them" (*J* IX:156–57).[3]

In this retrospective account, Thoreau announces a new form of attention and a new way of knowing. Moreover, as of 1850 he joins these new ways of thinking with a new way of acting. If he is to know his neighbors, he must visit them repeatedly. Thus Thoreau begins to take hundreds—thousands—of botanical excursions in the neighborhood of Concord. From 1850 to his death in 1862, Thoreau devotes himself to "observing when plants first blossomed and leafed," and he "follow[s] it up early and late, far and near, several years in succession, running to different sides of the town and into the neighboring towns, often between twenty and thirty miles in a day." In addition to these surveys, his walks lead to his encounters with specific plants, "four or five miles distant," and he visits them as much as "half a dozen times within a fortnight" (*J* IX:158). The excursions are both expansive and focused, and the evidence from the Journal suggests that his knowledge combines the scientific vocabulary of plant morphology and taxonomy with the philosophical, literary, and ethical dimensions promoted by Michael Marder as forms of plant-thinking.

This revolution in thinking is clearly the work of "several years in succession," a project that requires disciplined practices of thought and action. The Journal itself is a record of the work as well as the fundamental means of joining thought and action through the act of writing. On 16 November 1850, at the beginning of his botanical projects, Thoreau writes an important entry on his Journal and its role in the work he is undertaking:

> My Journal should be the record of my love. I would write in it only of the things I love. My affection for any aspect of the world. What I love to think of. I have no more distinctness or pointedness in my yearnings than an expanding bud—which does indeed point to flower & fruit to summer & autumn—but is aware of the warm sun and spring influence only. I feel ripe for something yet do nothing—can't discover what that thing is. I feel fertile merely. It is seed time with me— I have lain fallow long enough. (*PJ* 3:143–44)

The key to this passage is the extended metaphor of "yearnings," a kind of love distinct from human desire. Unlike the love of human beings, but very like an "expanding bud," Thoreau seeks a kind of awareness *as a plant*, pointing to some fruition that the writer himself cannot yet imagine. The botanical language gives Thoreau a way of figuring his present situation, an imaginative fertility that seeks a way to grow and develop into its fullest form, to "point to flower & fruit to summer & autumn." The passage may not present a fully realized version of plant-thinking, but it is an important early step in that direction. By imagining himself as a plant, Thoreau begins to imagine what a plant's noncognitive awareness might be, how it might yearn and expand, even though those processes do not necessarily conceive or achieve a conscious goal. Perhaps most salient of all, this imaginative cross-fertilization is both framed and enabled by the act of writing the Journal. By making a "record" of his love, Thoreau's love reaches many things and thoughts; it touches "any aspect of the world." Already, in the late fall of 1850, it fastens tenaciously upon plants as a form of imaginative being.

The year 1851 marks his complete botanical turn. Especially in the spring and summer of that year, the work of plant-thinking proliferates in several ways and directions. As the year continues, Thoreau gains skill in identifying plants, in describing them in detail, in learning their habits and habitat, and, sometimes, in perceiving their relation to human beings. He works at using botanical sources to aid his identifications and give him more and more genuine knowledge in the field, recognizing and discriminating between closely related species. He also maintains his aesthetic eye, but this is complicated by the other ways of knowing and responding that become important over the course of the year.[4]

On 13 February 1851, Thoreau observes the skunk cabbage flowering: "Examined by the botany All its parts—the first flower I have seen, the ictodes foetidum Saw in a warm muddy brook in Sudbury—quite open & exposed the skunk cabbage spathes above water— The tops of the spathes were frostbitten but the fruit sound— There was one partly expanded— The first flower of the season—for it is a flower—I doubt if there is a month without its flower" (PJ 3:191). This is a forerunner passage, markedly different from an earlier 1850 Journal entry in which the skunk cabbage leaves are, perhaps predictably, part of "a book which nature shall own as her own flower, her own leaves" (3:62). In 1851, rather than employing the tired trope of the "book of Nature," Thoreau marks the first flowering plant of the year. He is also thinking about larger patterns of flowering by the month. He uses a botany manual to examine all the parts of the plant. He cites the plant's scientific name. Jacob Bigelow's third edition of *Florula Bostoniensis* (1840) is most likely Thoreau's "botany" in this passage; Bigelow names skunk cabbage *Ictodes foetidus* and gives Thomas Nuttall's *Symplocarpus foetidus* as a synonym (FB 61–62). Bigelow's manual follows the Linnaean "artificial" system of arrangement, though in the manual's 1840 preface

Bigelow appears well aware of the new "natural" system of arranging the plants. Thoreau is right to be drawn to the botanical manual, for Bigelow's descriptions of plants are detailed, precise, and clear. In addition, the third edition contains a useful glossary of botanical terms.

Bigelow and Harvard botanist Asa Gray are Thoreau's two main authorities for identifying species. On 20 May 1851, he quotes "the botanist, Gray," on organs of vegetation and reproduction, and he develops an analogy to the human body/mind (*PJ* 3:224–28). The rootedness of plants makes Thoreau conjecture that "no thought but is connected as strictly as a flower, with the earth," and that "if our light & air seeking tendencies extend too widely for our original root or stem we must send downward new roots to ally us to the earth" (227). This earthy example of plant-thinking goes far beyond Gray's observation of plant physiology. Bigelow is cited in 1851 in regard to medicinal plants, suggesting that Thoreau is consulting Bigelow's textbook *American Medical Botany* (1817–20), but he is certainly using *Florula Bostoniensis,* too. Through the spring and summer of 1851, Thoreau cites Gray, Bigelow, and the botanist George Emerson for Latin names of plants he is identifying. He reflects repeatedly on the relations of humans to plants, with medicine as a major focus. On 29 May 1851, for example, he makes a long entry quoting Bigelow on medicinal plants (238–39). In June, Thoreau comes back to Bigelow on other medicinal plants, once again citing *American Medical Botany.* Especially in the first weeks of his botanical studies, he repeatedly quotes and cites Bigelow's *Florula Bostoniensis.*[5]

As his knowledge rapidly grows, Thoreau begins to outgrow his early guide. On 25 August 1851, he proudly notes, "I now know all the Rhus genus in Bigelow" (*PJ* 4:14). In Bigelow's Linnaean system, the *Rhus* genus is part of class 5, order 1, listed as number 128. The index of the *Florula Bostoniensis* sends the reader to page 118, where we find the *Rhus* genus, with five species described in detail (*FB* 118–20). On 21 August 1851, Thoreau quotes Bigelow on the spikes of blue vervain, then adds his own remarks about how the plant blossoms from base to summit and how this phenomenon gives "the true time of the season" (*PJ* 4:4–5) Despite his dedication to the task of identification, Thoreau can assert that "the poet is more in the air than the naturalist[,] though they may walk side by side" (4:6). Even as a practicing botanist, moreover, Thoreau sees limitations in Bigelow's use of the Linnaean artificial system of classification. He notes several flowers on 25 August and says that he keys unknowns to class 5, order 1, or to class 14, or to class 8, order 1 (4:13). Then he returns on 28 August to the "pretty little blue flower in the Heywood Brook—Class V ord 1" and writes out a careful description, likening it to the forget-me-not (4:15–16). The difficulty is that, in Bigelow's system, class and order include many genera, without providing a clear path toward identifying the correct genus or arriving at the correct species. This is not the kind of dichotomous key that modern manuals provide, and there is no clear way in which the sequence of families or genera suggest diagnostic similarities in structure. The class/order system is descriptive but general, following Linnaeus's method of using the number of stamens and pistils as a means of determining class and order, respectively. In other words, class 5 is "Five stamens," and order 1 is "One pistil" (*FB* 64). This classification system can lead a user to several suborders and a total of thirty-six genera. In the case of the "pretty little blue flower," however, the identification remains up in the air.[6]

Asa Gray's *Manual of Botany* gradually takes over as Thoreau's authoritative guide for identifying plants. Although never personally acquainted or in any way closely related, Gray and Thoreau share a scientific approach to identifying plants in the field and collecting specimens for an herbarium. As Fisher Professor of Natural History at Harvard University and curator of the Harvard Botanical Garden, Gray was the quintessential academic botanist, both as a researcher and as an academic lecturer. Gray also exerted a broad botanical influence through a series of accessible and popular textbooks. Works such as the *Botanical Text-book* (1842), which went through six editions by 1879, and the *Manual of Botany* (1848), which was revised in 1856 and reached a sixth edition in 1889, established Gray's reputation as the preeminent professional American botanist of his time.[7]

Gray's *Botanical Text-book* is an authoritative introduction to the science of botany as known in the middle of the nineteenth century. The writing is careful, precise, and dry. Gray takes the student through ten long chapters detailing the morphology of flowering plants. The book also delivers, in effect, a systematic vocabulary lesson, teaching the terms for describing and understanding the physical structure of the vegetative and reproductive parts of any flowering plant. Illustrations and examples abound, all discussed in great detail. The possibilities to be met by any field botanist anywhere in the world would appear to be included in the account. General patterns and tendencies of growth, form, and function dominate the book, but Gray treats exceptional cases of all kinds. In addition, the book explains how the science of botany has developed historically, focusing especially on the development of taxonomy and the terms used to classify plants systematically. Gray discusses Linnaeus, his distinction between the artificial and natural systems of classification, the artificial system Linnaeus devised, and the later natural systems developed by his successors, most notably Antoine Laurent de Jussieu and Augustin de Candolle, comparing their usefulness for present-day students. He presents the rules for nomenclature and description, advice on collecting and preserving specimens in an herbarium, and ends with a glossary of botanical terms that runs over seventy pages.[8]

Gray's descriptions of individual species in the *Manual of Botany* are not as full and detailed as Bigelow's in *Florula Bostoniensis,* but Gray includes many more species and covers a much broader geographical area than Bigelow. Moreover, brevity of description is a virtue. As Gray notes in his discussion of descriptive methods in the *Botanical Text-book,* Linnaeus recommends Latin descriptions of no more than twelve words. More important, the *Manual* gives the user an opportunity to follow an analytical structural key, focusing especially on the reproductive parts of the plant, in order to determine the family (called "order" in the nineteenth-century nomenclature). From there, the synopsis of orders takes a user to the description of the order and to the genera and species belonging to it. In the case of the *Rhus* genus, for instance, we learn that it belongs to the *Anacardiaceae,* or cashew family, and that there are seven species identified by Gray as belonging to the genus. While modern taxonomies have altered some of the details of Gray's nomenclature and classifications, his research remains the foundation for modern systematic botany.

Bigelow and Gray combine to train Thoreau in the observation of vegetative and reproductive parts of plants, the technical vocabulary for describing those parts precisely, and patterns that establish constellations of characteristics for plant families, genera, and species. The main credit for Thoreau's rapid progress in botany, however, should go to Thoreau himself. Through his disciplined practice of repeated excursions, with the observations he makes, the notes he takes, the specimens he collects, and the studies he prosecutes in his reading and writing, Thoreau becomes a proficient field botanist, as he notes in the Journal on 4 December 1856. Four rough approximations give an idea of Thoreau's accomplishments. The index to the 1906 Journal lists over 650 entries for plants, most by scientific name. The Harvard University Herbarium lists some 1295 sheets with Thoreau as collector. Following Walter Harding, Laura Dassow Walls gives the figures at 800 species of plants and over 1000 herbarium specimens. Most recently, Ray Angelo has accounted for over 900 species and at least 1500 specimens in the herbarium. These numbers suggest the degree to which Thoreau dedicated himself to the study of plants, even though no number is exact.[9]

One intellectual factor in his rapid progress is Thoreau's embrace of botanical vocabulary, especially relating to the morphology of the plant. He appreciates how "copious & precise the botanical language to describe the leaves, as well as the other parts of a plant. Botany is worth studying if only for the precision of its terms—to learn the value of words & of system" (*PJ* 3:382). His fear is that science narrows his vision "to the field of the microscope. . . . I count some parts, & say, 'I know'" (3:380). Yet botany provides an adequate language for describing the abundance of the plants Thoreau counts as neighbors, and a practiced user of a botanical manual can in fact take parts and find wholes. For Thoreau, the wholes are more than species, genera, or families of plants. On 20 August 1851, he exclaims, "The precision and copiousness of botanical language applied to the description of moral qualities!" (3:382). As a literary artist, of course, Thoreau immediately develops a skill in employing the botanical language of description. As a transcendentalist thinker, he applies botanical language to moral qualities. Botany gives him a new terminology for his imagination, one that is fundamental to plant-thinking. While the metaphorical value of botanical language is undeniable, the terminology in addition allows Thoreau to explore new ways of thinking about the relationship of plants to human beings. Metaphor can become more than mere metaphor.

In addition to these growing skills, Thoreau learns to observe larger patterns of plant growth and reproduction. Phenology—the observation of seasonal and cyclic phenomena such as first flowerings, arrival and departure of migratory birds, nesting and fledging, first fruiting, the turn of color in autumn leaves, and manifold other occurrences—becomes part of Thoreau's regular thought process in the spring and summer of 1851. During this time he starts noting first flowers, a practice that he develops in multiple ways and directions over the next decade. In such moments, the prose of the Journal takes on a tone of palpable excitement. On 29 and 30 June 1851, for instance, Thoreau ranges about the fields and roads of Concord, filling some eight pages of the Journal with observations of "old-country methods of farming resources" and descriptions of thirty-four separate species of flowering plants. His notes are both terse and evocative, emphasizing the act of observation as well as the plant observed: "I look down on rays of prunella by the road sides now— The panicled or privet Andromeda with its fruit-like white flowers— Swamp-pink I see for the first time this season" (*PJ* 3:278). The entries mix botanical observations with reflections on other topics. He mentions "the panicled cornel a low shrub in blossom by wall sides now" and then immediately follows the sentence fragment with a paragraph on the "*hypaethral* character" of his first book, *A Week on the Concord and Merrimac Rivers* (3:279). Like a bookend, after the literary paragraph the remark follows: "The potatoes are beginning to blossom" (279). These seemingly idiosyncratic, provocative combinations create a poetics of observation and appreciation. The actions of the eye join with the actions of the perceiving and reflecting mind.

In this same period, the botanical turn leads Thoreau to prolific writing. He fills 38 Journal pages in May 1851; 56 pages in June; 46 pages in July; 73 in August; 102 pages in September; 48 in October; 48 in November; and 38 in December.[10] The number of pages correlates roughly with the number of plants that are leafing out, budding, blossoming, and fruiting. Even more tellingly, Thoreau registers a rising enthusiasm in the spring and summer, as the plants turn out in their numbers and colors and identities. The excitement of the botanical excursions around Concord in turn leads Thoreau to thoughts about famous botanists and other travelers, such as Michaux, Bigelow, Gray, Darwin, and Humboldt. These writers lend authority to Thoreau's botany in the field and contribute to his reflections and writing at home, after the day's excursion. The Journal becomes a layering of these concerns, a record of multiple aspects of the writer's love. As a result, Thoreau expresses considerable concern with his own writing; his self-awareness of his work as a writer is a transcendental aspect of his own self-development and self-culture. What will he do with this material? What kind of writer is he? What kind of writer will he become? For Thoreau, the answers lie in the dynamic, twinned activities of botany and writing.

Typically, as 1851 moves into summer and fall, Thoreau heads the journal entry with the itinerary of the day's excursion (often an afternoon hike, although he also roams in the mornings and nights); the entry then gives observations, descriptions, and identifications of plants, as well as reflections on other topics such as local history, human neighbors, birds and other wildlife, and agricultural practices. When he discusses plants, especially in the early years, Thoreau sometimes questions himself about the identification. As we have seen, he is very proud when he can say, for instance, "I now know all the Rhus genus in Bigelow" (25 August 1851). At other times, he cautions himself about putting too much emphasis on acquiring scientific knowledge, even if it seems to be a source of pride. In the long entry of

19 August 1851, less than a week before he announces his botanical accomplishment in the *Rhus* genus, after recording several identifications of plants and reflecting on the scientific names of birds and flowers, he considers, in a more emotional, intimate way, the state of his own knowledge: "I fear that the character of my knowledge is from year to year becoming more distinct & scientific— That in exchange for views as wide as heaven's cope I am being narrowed down to the field of the microscope— I see details not wholes nor the shadow of the whole. I count some parts, & say, 'I know.' The cricket's chirp now fills the air in dry fields near pine woods" (*PJ* 3:380). More than a simple record, Thoreau's journaling is itself a journey, a series of waymarks along the path he is discovering. The goal, in terms used by Laura Dassow Walls, is the "empirical holism" that creates wholes out of parts. But, at this point of rigorous self-examination, Thoreau cannot discern even the "shadow of the whole."[11]

The month of July 1851 may be especially helpful in charting the writer's botanical path. In manifold ways, Thoreau is at a turning point in his career and life. He reflects on these very matters in clear terms during July. On 19 July, for example, he notes, "Here I am 34 years old, and yet my life is almost wholly unexpanded. How much is in the germ!" (*PJ* 3:313). As the reflection continues, he entertains the idea that his spirit might develop seasonally, but not in direct correlation to the annual seasons of nature: "If my curve is large—why bend it to a smaller circle? My spirit's unfolding observes not the pace of nature" (3:313). Later that afternoon, the entry headed "2 Pm" features a host of flowering plants that foretell a shift from summer to autumn. Without naming the species, Thoreau meets "the first orange flower of autumn— What means this doubly torrid—this Bengal tint— Yellow took sun enough—but this is the fruit of a dogday sun. The year has but just produced it" (3:314). This passage does not record the botanist's sense of an early flowering or some phenological change; instead, the color of the flower suggests autumn, so that the flower is the figurative "fruit of a dogday sun." The Journal is recording a season in the writer's spirit. The season, moreover, is ambiguous. Even though many of the wildflowers are now past their prime, they still fill the air with their perfume and scatter their petals over the leaves of "neighboring plants." Thoreau finds himself in a liminal season, where the seasons are more than a relentlessly linear march of hours and days. And, for the writer, whose life and work are still "in the germ"—what season is this for him? The weaving of plants and the writer's imagination is a prime example of plant-thinking, in Marder's sense of an encounter. For Thoreau, moreover, the encounter is characteristic in that it raises more questions than it can possibly reduce to a single answer.

By developing his skills of observation, Thoreau hones his ability to connect the plants to his own life and work. On 23 July, he ends a long entry on the play of mind and external influences ("the ovipositors plant their seeds in me I am fly blown with thought") by again questioning his propensity for close observation, the very skill he admires in travelers like Darwin and Humboldt: "But this habit of close observation—In Humboldt—Darwin & others. Is it to be kept up long—this science—Do not tread on the heels of your experience Be impressed without making a minute of it. Poetry puts an interval between the impression & the expression—waits till the seed germinates naturally" (*PJ* 3:330, 331). As in the entry of 19 July, the figure of the germinating seed gives Thoreau a way to examine himself—in this case, his developing habit of closely observing plants and recording a "minute" of his minute impressions. The self-examination springs from the practice of botanical observations and notations, the developing of the botanical vocabulary in detail, and then the application of these practices to his own imaginative life. This is clearly a practice of plant-thinking in several forms and dimensions. Without the practices of botany, Thoreau would not arrive at the image of writing as a seed that germinates naturally. Nor would he arrive at thinking about his own writing in ways that evoke the life of plants. His plant-thinking also engages with the issue of plants' temporality, for his botanical practices are not the fruit of one month or one year alone. Over the next decade, Thoreau repeatedly uses the suite of practices associated with botanical excursions to refresh and enliven his imaginative engagement with the world around him. If poetry puts an interval between impression and expression, the Journal is the place in which Thoreau manages to find and exploit that interval.

Given this journey, both in the near-daily excursions and the Journal entries, it is not surprising that temporal succession would become a meaningful aspect of Thoreau's plant-thinking. While the passing of time registered in plants is closely related to phenology, it is distinct from it. The first leafing, or first flowering, or first seeds, or the natural germination of a seed—such markings of the passage of time and the order of seasons express the succession of plants from one stage to another. This is not only the sense of succession as a large-scale ecological process, but also the sense of "seasons" succeeding one another, in large ways or in small. A "season," in this sense, is more than the four traditional seasons of a calendar year.

For example, a specific day's observation might become a marker of "the time of the apple," or, as in the Journal of 19 July 1851, a color may mark a turn to autumn, or to the idea of autumn, even though the season itself is weeks or months away. Because Thoreau is thinking seasonally as he makes the botanical excursions and observations, his thinking often moves beyond the botanical observation to encompass his own life as a writer. In the long entry for 23 July 1851, he considers switching his habitual afternoon excursions for morning walks, so that he might write better in the afternoon: "It would be like a new season to me," he writes, "& the novelty of it inspire me" (*PJ* 3:329). While these examples do not all signify in the same way, they show that the idea of the seasons is a major part of Thoreau's plant-thinking, one that is more complicated and multifaceted than the simple passage of the four seasons in an annual cycle.[12]

The relationship between plant life and human aspirations is a repeated motif in the 1851 Journal. On 30 August, in extended thoughts on Norway cinquefoil (*Potentilla Norvegica*), Thoreau focuses on how the five leaves of the calyx close over the seeds to protect them, conjecturing that this is "evidence of forethought, this simple *reflection* in a double sense of the term . . . as if it said to me Even I am doing my appointed work in this world faithfully" (*PJ* 4:18). He uses the terms "faithfully" and "faithful" to describe this recurring action and asks of himself to act as faithfully in his own "appointed work." He sees the action of the plant as "one door closed, of the closing Year" (18). The change to fruiting time, in Thoreau's plant-thinking, is "not an obscure—but essential part" that the plant plays "in the revolution of the seasons" (18). He insists, moreover, that "the fall of each humblest flower marks the annual period of some phase of human life . . . experience" (19). In a recursive observation, he asserts that he sees the flower's change only when he sees in it "the symbol of my own change" (19).

It is vital to note the ways in which Thoreau engages in plant-thinking in this passage. The point is not the usual trope of recognizing a flower as a figure for a human being or the human spirit; rather, it marks Thoreau's way of preparing himself to recognize the changes in flowering plants and other parts of the world he meets on his botanical excursions. If the observer is prepared, he or she may find a corresponding fact or event to answer to a yet-unformed question about his or her own changes. The observer's preparation, the key to the encounter with "the fall of each humblest flower," is a signal development of Thoreau's plant-thinking. Even though preparation is the key, the process steps away from human "pointedness" toward a productive encounter with changes in the temporal life of plants. "When I experience this then the flower appears to me" (*PJ* 4:19). This paradoxical statement reverses the usual order of perception and meaning. By preparing himself for the encounter, Thoreau experiences the reversal as a form of recognition, when the flower appears to him in all its significance, as a "symbol of my own change." In this kind of plant-thinking, the human being prepares for the encounter by approaching the state of a plant in a period of transition. The protectiveness of the Norway cinquefoil becomes an encounter because Thoreau prepares himself imaginatively for such a significant moment. Clearly, this creation of meaning does not occur because Thoreau wishes it or expects it to occur; the creation of meaning occurs because he engages in plant-thinking as a form of imaginative discipline and preparation. Then the flower appears to him.

The practices of plant-thinking include observation, notation, reflection, writing, reading and study in scientific texts. Plant-thinking occurs in the kind of imaginative excursion that takes place in all manner of places, from an attic bedroom to a neighboring swamp. In the narrative of a moonlight excursion, recorded on 11 June 1851, Thoreau creates the experience of walking from the Deep Cut at Walden Pond to Fair Haven Pond by the railroad tracks, then returning home by way of Potter's pasture and the Sudbury road (*PJ* 3:249–59).[13] The mix of present and past tense intensifies the sense of actions and perceptions taking place as one reads the account, and the strategic use of deictics like "now" adds to the effect of immediacy. As the narrative proceeds, Thoreau intersperses reflections, as if the thoughts are arising directly from the experiences he narrates. For example, he hears a "night singing bird breaking out as in his dreams," and then he writes, "Our spiritual side takes a more distinct form, like our shadow which we see accompanying us" (3:252). A bit later, after noting, "By night no flowers—at least no variety of colors," he returns to the image of his shadow, like "a 2nd person—a certain black companion bordering on the imp" (252–53).

Suddenly he opens the narrative to another, unexpected insight: "No one, to my knowledge, has observed the minute differences in the seasons— Hardly two nights are alike— The rocks do not feel warm to-night, for the air is warmest—nor does the sand particularly. A Book of the seasons— each page of which should be written in its own season & out of doors or in its own locality wherever it may be—" (253). Such a "Book of the seasons" broadens the definition of "season" in every way, so that each page is written in a season of its own and in a locality of its own. That "Book of the seasons" is clearly an ideal for the Journal, and, as Thoreau concludes the narrative of the moonlight walk, he notes that the return by the Sudbury road takes the walker out of place and out of season: "You catch yourself walking merely The road leads your steps & thoughts alike to the town— You see only the path & your thoughts wander from the objects which are presented to your senses— You are no longer in place" (253). By writing the entry the following day, Thoreau uses narrative to recreate the experience of the moonlight walk, its objects and thoughts. Through the writing, he recovers a sense of locality and season, putting himself "in place." In an intriguing development, the entry moves from the moonlight excursion to a set of summaries and excerpts from Charles Darwin's *Journal of Researches* (later titled *Voyage of the Beagle*), as if Thoreau's adventure somehow reflects upon the "Voyage of a Naturalist round the World" (253–59).

TRAVELING IN YOUR OWN COUNTRY

Thoreau's repeated encounters with plants stem from the botanical excursions he makes "in place," fastening himself and his love to the locality and seasons of Concord. The encounters help him develop, in turn, two related and fundamental figures in the Journal: the "traveller" and "the walker." The figures become closely tied to one another through writing projects he undertakes in the 1850s, such as "Walking," *Cape Cod*, and *The Maine Woods*. They reappear significantly in late works like *Wild Fruits*. On 2 July 1851, Thoreau recalls a humorous episode in which he is awakened around one o'clock in the morning by two loud-talking travelers, "whiling away the night with loud discourse" (*PJ* 3:283). Even though he might be peeved at such behavior, Thoreau takes it in good form: "A Traveller! I love his title A Traveller is to be reverenced as such— His profession is the best symbol of our life. Going from-toward— It is the history of every one of us. I am interested in those that travel in the night" (283). A few weeks before, on 12 June, he had followed the long entry of 11 June, full of those passages from Darwin's *Voyage of the Beagle* and his own idea of "a Book of the seasons," with a thoughtful admonishment: "There would be this advantage in travelling in your own country even in your own neighborhood, that you would be so thoroughly prepared to understand what you saw— You would make fewer travellers' mistakes. Is not he hospitable who entertains thoughts?" (3:259).

As Thoreau takes his botanical turn in 1851, he directly considers the question of locality in developing his projects. On 6 August 1851, he writes a long reflection on the question of travel: "A man must generally get away some hundreds or thousands of miles from home before he can be said to begin his travels— Why not begin his travels at home—!" (3:356). After all, the traveler abroad carries the "principal distinction . . . that he is one who knows less about a country than a native" (357). For Thoreau, "it takes a man of genius to travel in his own country—in his native village—to make any progress between his door & his gate" (357). His thoughts run to how he learns the most in his local circumstances, not in the novelties of faraway places. "I wish to get a clearer notion of what I have already some inkling" (357). In the same entry, however, he cites two of his favorite explorers, William Bartram and F. A. Michaux, as authorities in both travel and botany. In 1850 and 1851, he makes extensive notes on Alexander von Humboldt's *Aspects of Nature* and, as we have just seen, Charles Darwin's travel narrative, *Voyage of the Beagle*. Indeed, the botanical authorities Thoreau seems most drawn to in his early studies are travelers. André Michaux and his son François André Michaux, John and William Bartram, Peter Kalm—all these botanists were intrepid explorers of lands unknown to white Europeans.

On 12 August 1851, Thoreau finds himself in the middle of the night, under a full moon, sitting on the railroad sleepers, wide awake to the possibilities of the hour. "It does not concern men who are asleep in their beds," he writes, "but it is very important to the traveller whether the moon shines bright & unobstructed or is obscured by clouds. It is not easy to realize the serene joy of all the earth when the moon commences to shine unobstructedly unless you have often been a traveller by night" (3:363). The insight of the traveler depends on the repeated experience of being "a traveller by night," for only through the discipline of repeating journeys will the traveler see the unobstructed moon and realize "the serene joy of all the earth." Though limited to the local place, the traveler's vision is unobstructed by any limitation. A week later, however, in the long entry of 19 August, Thoreau is clearly worried about the limits on his travels: "How many things concur to keep a man at home, to prevent his yielding to his inclination to wander" (376). Unlike the migrating birds, human beings need a home. "As travellers go round the world and report natural objects & phenomena— so faithfully let another stay at home & report the phenomena of his own life. Catalogue stars—those thoughts whose orbits are as rarely calculated as comets It matters not whether they visit my mind or yours—whether the meteor falls in my field or in yours—only that it come from heaven" (377). In this essential individual perception, Thoreau proposes to create "a meteorological journal of the mind— You shall observe what occurs in your latitude, I in mine" (377). The key to such faithful reporting of phenomena, whether physical or mental, seems to be the repeated action of traveling or walking, even close to home: "Methinks that the moment my legs begin to move, my thoughts begin to flow" (378). In such repeated action, moreover, the walker/writer moves relentlessly, the way that nature moves: "The seasons do not cease a moment to revolve and therefore nature rests no longer at her culminating point than at any other. If you are not out at the right instant, the summer may go by & you not see it. How much of the year is spring and fall—how little can be called summer! The grass is no sooner grown than it begins to wither" (379). For the traveler at home, time is fleeting, as are all the phenomena rushing past us.

Two long, reflective entries for 7 September 1851 summarize much of the development we see in Thoreau's botanical studies and the ways in which plant-thinking takes on a central role in the Journal. Echoing the earlier entry of 16 November 1850, he describes himself as "uncommonly prepared for *some* literary work" but unsure what work that might be (*PJ* 4:50). He feels "that the juices of the fruits which I have eaten the melons & apples have ascended to my brain—& are stimulating it. They give me a heady force" (51). These ecstatic states "appear to yield so little fruit," but, in "calmer seasons, when our talent is active, the memory of those rarer moods comes to color our picture & is the permanent paint pot as it were into which we dip our brush" (51–52). This is Thoreau at his most faithful, certain that "no life or experience goes unreported" (52). In developing the line of thought, he sees clearly that "it is with flowers I would deal. Where is the flower there is the honey—which is perchance the nectareous portion of the fruit—there is to be the fruit—& no doubt flowers are thus colored & painted—to attract the bee" (53). The reproductive stages of plants give Thoreau confidence in the productive stages of his imagination. Later, narrating an excursion to Conantum, he sees industrious men raking cranberries and asks if the scene "shall reprove my idleness." His answer is to ask demanding questions of his own: "Can I not go over those same meadows after them & rake still more valuable fruits. Can I not rake with my mind? Can I not rake a thought perchance which shall be worth a bushel of cranber?—" (59). For Thoreau, the dialogue between plants and the traveler's imagination produces discoveries, and "the nearer home the deeper" (54).

Thoreau relates the lesson of the botanical turn and of repeated botanical excursions in an utterly charming entry of 31 May 1853, the story of being shown the "beautiful purple azalea or pinxter flower," called *Azalea nudiflora* by Linnaeus, *Rhododendron nudiflorum* by Gray. In the first part of the entry, Thoreau already sees clearly the lesson of the story—that some incidents in his life are "far more allegorical than actual," partaking more of myth than history. The experience of such an allegory is "quite in harmony with my subjective philosophy," and the allegory shows that "the boundaries of the actual are no more fixed & rigid—than the elasticity of our imaginations" (*PJ* 6:162). Like the story of the Norway cinquefoil, this tale includes "just such events as my imagination prepares me for—no matter how incredible" (162). For Thoreau, the pinxter-flower is an exotic, too rare and beautiful to have a place in his thoughts, but then it is suddenly "found in our immediate neighborhood," a fact that the writer finds "very suggestive."

On the evening of 30 May, his sister Sophia brings home a "single flower without twig or leaf" from a neighbor's home. This starts the writer's search. Thoreau follows a chain of flowers and neighbors to the hunter/trapper George Melvin, "sitting in the shade bare headed at his back door— He had a large pail ful of the azalea recently plucked—in the shade behind his house. which he said he was going to carry to town at evening" (164). Melvin has been out fishing that morning, and it appears that he has also been drinking all day. At first, Melvin won't tell where the flower grows, but Thoreau recalls having seen Melvin weeks before, "on Wheelers' land beyond the [Assabet] river," and he suggests that he will find the flower there, with or without Melvin's help. Finally, Thoreau uses as much leverage as he can muster to get the secret out of the reluctant Melvin:

> Well I told him he had better tell me where it was— I was a botanist & ought to know But he thought I couldn't possibly find it by his directions—, I told him he'd better tell me & have the glory of it—for I should surely find it if he didn't— I'd got a clue to it & shouldn't give it up— I should go over the river for it I could smell it a good way you know.— He thought I could smell it half a mile—& he wondered however that I hadn't stumbled on it—or Channing— Channing he said came close by it once—when it was in flower. (*PJ* 6:164–65)

The style mixes a homespun folk tale with mythic, allegorical tones, and Thoreau accentuates the mixture with a lofty, ironic humor. After a seemingly long delay, Thoreau accompanies Melvin and his dog across the river in his boat, and Melvin shows the botanist where the *Azalea nudiflora* grows. Thoreau offers to pay him for his trouble, but Melvin refuses: "He had just as leif I'd know as not. He thought it first came out last Wednesday on the 25th" (166). Thoreau follows the episode by noting that neither of his botanical guides, Gray and Bigelow, says anything about the plant's *"clamminess."* He then describes the shrub lovingly and at length, noting that the *lack* of "clamminess" accounts for the lack of insects around the flowers (166).

The country humor of this story, well known to Thoreau scholars and reprinted quite often, depends especially on Thoreau as the relentless "botanist," pursuing the flower from neighbor to neighbor, braving the drunken Melvin's hesitations, and using every persuasion—even glory—to gain the secret of the flower's location. The story does indeed illustrate the allegory of an elastic reality, more fluid and flowing than the botanist's knowledge. Only if the botanical traveler unfixes his imagination, making it as elastic as reality itself, can he be prepared to meet facts that are "lifted quite above the level of the actual" (162). The story also illustrates the knowledge that resides in the local neighborhood, among the botanical explorer's neighbors, available to anyone who has eyes to see, ears to hear, and a nose to smell. George Melvin emerges as a generous, innocent lover of the plant and its flowers—perhaps even the secret hero of the myth, an allegory of the selfless Hunter.

As we shall see in the following chapters, Thoreau takes the botanical turn in order to, as he puts it, "know my neighbors, if possible—to get a little nearer to them" (*J* IX:157). In addition, his botanical excursions enable him to get a little nearer to his human neighbors. Whether he is traveling near at home or farther afield, Thoreau learns and relearns the lesson of the pinxter-flower: "The boundaries of the actual are no more fixed and rigid than the elasticity of our imaginations." In *The Maine Woods, Cape Cod,* and *Walden,* Thoreau's botany productively tests the boundaries of the actual and the elasticity of the imagination.

PART I

Reconstructing the Botanical Excursions

CHAPTER 1

The Two Botanical Excursions
of *The Maine Woods*

From the beginning of Thoreau's career to the very end, the excursion is a prominent subgenre of travel writing and narrative nonfiction. One salient aspect of the excursion is the journey out and back. The narrative features the dynamic of crossing a boundary into the new and unknown, then returning home to the familiar or newly unfamiliar place of origin. In his first-person narratives, Thoreau is preeminently a traveler, even in Concord. In the opening paragraph of *Cape Cod,* published posthumously in 1865, he writes, "I have been accustomed to make excursions to the ponds within ten miles of Concord, but latterly I have extended my excursions to the sea-shore" (*CC* 3). The "excursion" can mean a short day-hike or walk, or it can mean a longer trip to a distant place. As a nonfiction narrative, the excursion can be a relatively short essay or a multichapter, book-length story. *A Week on the Concord and Merrimac Rivers* (1849), Thoreau's first book, is an excursion; the essay "A Winter Walk" (1843) is an excursion.

The first of four posthumous volumes edited by Thoreau's sister Sophia, *Excursions* (1863) is a loosely chronological miscellany of essays that may owe its title and some of its contents to Emerson.[1] The thematic thread connecting the essays seems to be natural history. The early excursion essays from the 1840s, like "A Winter Walk" and "A Walk to Wachusett," are charming narratives, but they are written before Thoreau makes his botanical turn. The review essay "Natural History of Massachusetts," originally commissioned by Emerson for publication in the *Dial,* is not an excursion and has little to say about botany, noting that the state reports "on Herbaceous Plants and Birds cannot be of much value, as long as Bigelow and Nuttall are accessible" (*Exc* 27). The absence or minimal presence of plants is as conspicuous as the prominence given to them. The longest travel narrative in the collection,

"A Yankee in Canada," shows early observations of plants and trees in the landscape, but Thoreau's 1850 excursion to Montreal and Quebec features most prominently the walk along the St. Lawrence River and visits to several waterfalls. Especially on the walks to the falls, Thoreau pauses to describe flowering plants and forest trees, but these descriptions are not foregrounded in the landscape or the writing. Thoreau also quotes the Swedish naturalist Peter Kalm in three places, but he uses the source for simple descriptive purposes. More interesting, in the final analysis, are the late natural history essays "Autumnal Tints" and "Wild Apples," published posthumously in the *Atlantic Monthly* in October and November 1862, but these are not really excursions at all, neither in the physical nor in the narrative sense. In fact, it seems clear that *Excursions* is an initial attempt to gather a chronological selection of Thoreau's previously published work into a book, using as a title the familiar term Thoreau employed for his walks and travels.

Unfinished at the time of Thoreau's death, *The Maine Woods* (1864) is the second posthumous volume, edited by his sister Sophia and published just a year after *Excursions*. The three parts are three separate narratives of excursions to Maine, in 1846, 1853, and 1857. The first two were published in magazines during Thoreau's lifetime; the last remained an unfinished manuscript.[2] The three narratives are important to this study because they offer a compendium of Thoreau's development as a botanist. The first narrative, "Ktaadn," is probably the best known and most read of the three. While it does show Thoreau's interest in the forests of Maine, it shows few signs of the botanical turn we have seen in the Journal of 1850–51.

In "Ktaadn," Thoreau gives the scientific name of only two plants: the mountain cranberry (*Vaccinium Vitis-Idaea*) and the black spruce (*Abies nigra*). He names several other trees, a few common shrubs like the blueberry, and a couple of other flowering plants. The vast, continuous evergreen forest is the most conspicuous botanical presence in the excursion. Thoreau presents the Maine forest as a dense, damp, often forbidding wilderness. He sees tremendous beauty and vitality in the woods as well. Early in the narrative, for example, he describes the Houlton Road (or Military Road) with the admiring eyes of a newcomer:

> The beauty of the road itself was remarkable. The various evergreens, many of which are rare with us—delicate and beautiful specimens of the larch, arbor-vitae, ball spruce, and fir-balsam, from a few inches to many feet in height, lined its sides, in some places like a long front yard, springing

up from the smooth grass-plots which uninterruptedly border it, and are made fertile by its wash; while it was but a step on either hand to the grim untrodden wilderness, whose tangled labyrinth of living, fallen, and decaying trees,—only the deer and moose, the bear and wolf, can easily penetrate. More perfect specimens than any front yard plot can show, grew there to grace the passage of the Houlton teams. (*MW* 11)

The repetition of the word "specimens" does not give this passage a scientific tone; instead, Thoreau seems to employ the language of horticulture to describe the landscaping effects of the roadside trees. The passage evokes the writer's pleasure in cultivated, civilized plantings, contrasting the orderly scene of the road with the "tangled labyrinth" of the "untrodden wilderness." Moreover, by depopulating the wilderness, Thoreau makes it the home of only wild mammals. The traveler's pleasure is heightened by the rarity of the trees in Concord, so that the "front yard plot" in Maine is a pleasing mixture of the familiar and unfamiliar.

A similar sense of both connection and contrast between the cultivated and the wild informs the final paragraphs of "Ktaadn." Thoreau ends with a five-paragraph coda, a self-contained essay that summarizes the lessons of the Maine woods. The most striking aspect of the Maine wilderness, he asserts, is "the continuousness of the forest, with fewer open intervals or glades than you had imagined" (*MW* 80). This unbroken continuousness creates the effects of the "grim and wild," "damp and intricate," "wet and miry," "savage and stern" wilderness. Only the lake prospects offer a "mild and civilizing" alternative to the string of dark doublets. The wildness marks the Maine woods as "not the artificial forests of an English king—a royal preserve merely"; instead, the continuous forest remains a preserve of nature and its laws, where "the aborigines have never been dispossessed, nor nature disforested" (80). This idealized vision leads Thoreau to figure the continuous forest as growing "ever more young," as being "exceedingly new," and as remaining "unsettled and unexplored" (81). The continuous forest becomes a metaphor for "that very America" visited centuries before by the old explorers; even the eastern coasts do not appear truly "discovered and settled" (82). Thoreau's admittedly naive vision of the Maine woods unsettles the continent and reinhabits it with continuous forests, abundant wildlife, and the Penobscot people. In the final sentence, he peers into the continuous forest to find that the "country is virtually unmapped and unexplored, and there still waves the virgin forest of the New World" (83).

While "Ktaadn" is certainly an excursion narrative, it is not a botanical excursion. That is, the forests of Maine are important to the narrative, but not central, and the encounter with the trees and plants does not underpin either the story of Thoreau's climb of Mt. Katahdin or his sense—both near the summit and again in the descent through the Burnt Lands—of a wild, untamable, vast, and inhuman reality. While the encounter does register the otherness of nature, Thoreau does not develop an image of *phusis,* nature, as a vital, dynamic process of continual change. Nor does he engage in the multiple practices of plant-thinking that we find in the Journal of 1851.

FORMS OF GEOGRAPHY IN THE APPENDIX

The seven-part appendix to *The Maine Woods* registers most readily Thoreau's fascination with botany and the geographical distribution of plants, and in that regard it signals most readily Thoreau's botanical turn of 1851. The first three parts—over two-thirds of the appendix—are devoted to plants. Especially in the 1857 trip, Thoreau locates the plants, both in geography and in terms of habitat. This effort at plant geography summarizes an entire process of observing, collecting, identifying, and describing, a process that the Journal of 1851 aligns with plant-thinking. Thoreau's botanical focus also echoes older naturalists like Carl Linnaeus, François André Michaux, and Frederick Pursh and pays tribute to his studies in Alexander von Humboldt, Charles Darwin, Asa Gray, and other naturalists. While Thoreau is working toward an intelligent sense of the geographical distribution of plants, in the 1850s this remains a work in progress.

However, the appendix clearly shows that by 1857 Thoreau's knowledge of botany, especially taxonomy, was both extensive and deep. According to section 3, "List of Plants," during the Maine excursions he identified 23 trees, 38 small trees and shrubs, 145 herbaceous plants / small shrubs, and 9 plants "of lower order" such as common grasses, horsetail, and clubmoss. The original draft of the "List of Plants" is a ten-page entry for the Journal of October 8, 1857. The published "List of Plants" synthesizes all three excursions to Maine, though most of the plants come from the two in 1853 and 1857. The revisions and expansions in the printed text further suggest that Thoreau revised the appendix before his death. Together with the narratives of "Chesuncook" and "The Allegash and East Branch," the appendix shows that the two later excursions were focused on Thoreau's deepening observations about trees, shrubs, and forbs. In addition, it provides some evidence that Thoreau was seeking to merge his botanical expertise with his growing appreciation for and knowledge of Native Americans in the Northeast. By 1857, he was making concrete advances in bringing the two ways of knowing into conversation with one another.[3]

In section 1, "Trees," Thoreau records nineteen species in terms of prevalence and location. For instance, he notes that fir and spruce form "very dense 'black growth,' especially on the upper parts of the rivers" (*MW* 298). Aspen, on the other hand, is "very common on burnt grounds," while alder "abounds everywhere along the muddy banks of rivers and lakes, and in swamps" (298–99). Less abundant than the fourteen common species are five species that grow only on islands or other "particular places," an observation that suggests Thoreau's knowledge of locally endemic species. Overall, he notes that the trees he records are "almost all peculiarly Northern trees, and found chiefly, if not solely, on mountains southward." Here we see Thoreau's understanding of the relation between latitude and altitude, an aspect of biogeography hypothesized by Alexander von Humboldt.[4]

In section 3, "Flowers and Shrubs" (*MW* 299–304), Thoreau organizes his remarks by habitat. Dispersion of wildflowers is not as great as in a "cleared and settled country," and Thoreau sees the wildflowers in Maine as "the pioneers of civilization." The implications of this metaphor are unstated, but the figure suggests that plants are somehow like white settlers, eventually transforming the Maine woods into a "cleared and settled country" like Massachusetts. Thoreau lists thirteen species that exist in the woods themselves, despite the lack of sunlight, and he proceeds to list flowering plants by location: river and lake shores (37 species, and 2 of "inferior orders"); in the water (4); in swamps (3); burned grounds (2); cliffs (5); on carries and old camps (15). The "underwoods" include small trees and shrubs such as moosewood, mountain maple, hobblebush, and American yew (4 species). Focusing on physical location, he notes the shrubs and small trees along river and lake shores (16); in swamps (6); on carries and in old camps (5); and on mountains (1). Finally, he observes some sixteen plants "*introduced* from Europe," most occurring in clearings, camps, and carries; he adds "about a dozen" plants that were "naturalized" by 1853, when he saw them on logging trails and roadsides (303–4). He does not name the "naturalized" plants, but they are likely some of the same species that he names as "introduced." In both cases, the plants have followed white settlement practices of clearing and building; Thoreau seems to be following a similar story in the way he describes the botany of Maine.

Thoreau's narrative of Euro-American settlement is not the only story of naturalized plants. Interestingly, one of the introduced plants Thoreau names is *Plantago major,* common plantain. An Indigenous name for the plant is "White Man's Footstep." In her groundbreaking book *Braiding Sweetgrass,* botanist and Native storyteller Robin Wall Kimmerer describes the arrival of this round-leafed, low-growing plant in North America: it came "with the first settlers and followed them everywhere they went. It trotted along paths through the woods, along wagon roads and railroads, like a faithful dog so as to be near them."[5] At first, says Kimmerer, the Native people distrusted the plant, which "came with so much trouble trailing behind," but ultimately they began to "learn about its gifts." She writes: "In spring it makes a good pot of greens, before summer heat turns the leaves tough. The people became glad for its constant presence when they learned that the leaves, when they are rolled or chewed to a poultice, make a fine first aid for cuts, burns, and especially insect bites. Every part of the plant is useful. Those tiny seeds are good medicine for digestion. The leaves can halt bleeding right away and heal wounds without infection" (214). Kimmerer notes that, while many "immigrant" plants do not make themselves welcome on a new continent, White Man's Footstep has now become an "honored member of the plant community"; it has "earned the name bestowed by botanists for plants that have become our own. Plantain is not indigenous but 'naturalized.' This is the same term we use for the foreign-born when they become citizens in our country" (214).

In an analogous and more general manner, Kimmerer writes, nonindigenous human beings can become "naturalized to place": "Being naturalized to place means to live as if this is the land that feeds you, as if these are the streams from which you drink, that build your body and fill your spirit. To become naturalized is to know that your ancestors lie in this ground. Here you will give your gifts and meet your responsibilities. To become naturalized is to live as if your children's future matters, to take care of the land as if our lives and the lives of all our relatives depend on it. Because they do" (214–15). This story of naturalizing, of "becoming indigenous to place," is no easy process. The common plantain takes some five hundred years to earn its place and reveal its gifts. For human beings, "to become naturalized" means committing oneself and one's culture to an ethic of care, both for the land and for its many inhabitants. The commitment reaches back into the past of our ancestors and into the future of our children; it creates a network of responsibilities and gifts.

Both sections 1 and 2 of the appendix are, in effect, short articles about the botany of the Maine woods, resembling in that regard the kind of article one finds in François André Michaux's *North American Silva,* a three-volume book that Thoreau knew well. Like articles in every domain, they open themselves to other stories, other ways of knowing. Section 3, on the other hand, a "List of Plants," is a long catalog of some 215 species of trees, shrubs, plants, and cryptogama, identified in 1853 and 1857, though it does include the 2 species he identified on the 1846 excursion. Thoreau repeatedly notes the specific locations where he saw or collected specimens of specific plants. Section 3 is divided into four subsections: 1) attain height of trees (23 by his count); 2) small trees and shrubs (38 by his count); 3) small shrubs and herbaceous plants (145 by his count); and 4) Of lower order (9). He notes the place observed and the relative abundance of individuals (common, abundant, uncommon, not abundant), and he includes occasional question marks for identifications. He places an asterisk on species that do not occur at all in the woods. For example, *Betula alba* (American white birch) occurs around Bangor only (*MW* 305).

In all three sections, Thoreau joins scientific names with common names, using them often as interchangeable terms. For very familiar plants, he uses a common name, often with the Latin binomial in parenthesis. For others, he gives the scientific name, in most cases taken from Asa Gray's *Manual,* and adds a common name in parenthesis, or he simply gives the scientific name. Occasionally he adds a comment on the plant and its habitat, or other qualities. For instance, about *Aster macrophyllus,* large-leaved aster, he remarks that "the whole plant [is] surprisingly fragrant, like a medicinal herb" and notes where he found the plant in 1853 and 1857 *(MW* 308).

Only in section 7, "A List of Indian Words" (*MW* 320–25), does Thoreau give Native American names for plants, and they are confined to eleven Penobscot words, two from 1853 and nine from 1857. Section 7 also includes Native names for birds, mammals, lakes, rivers, mountains, and other landforms. Curiously, it includes a separate list of place names taken from William Willis, "The Language of the Abnaquies, or Eastern Indians," but the first part of the section names personal contacts such as the guides Joe Attean and Joseph Polis in addition to scholarly sources.[6] Even though this section of the appendix indicates that Thoreau's knowledge of Native languages is slim, it approaches the three botanical sections in length. In the 1857 excursion, Penobscot language plays a more important role in *The Maine Woods* than the appendix at first indicates. In a similar way, the three botanical sections cannot fully represent Thoreau's botany, for his methods of detailed observation constitute an encounter with plants that combines scientific taxonomy with emotional, ethical, and aesthetic responses. Ultimately, the developing role of Thoreau's botany emerges from a reading of the botanical excursions of 1853 and 1857.

THE MOOSE, THE PINE TREE, THE INDIAN

The narrative of "Chesuncook" differs strikingly from the focus on mountain climbing in "Ktaadn," despite important echoes of the first excursion. The most significant new development is the way in which Thoreau foregrounds his botanical work on the trip. Like the experience on the Houlton Road, the road leading from Bangor to Moosehead Lake elicits Thoreau's exclamations: "It rained all this day and till the middle of the next forenoon, concealing the landscape almost entirely; but we had hardly got out of the streets of Bangor before I began to be exhilarated by the sight of the wild fir and spruce tops, and those of other primitive evergreens, peering through the mist in the horizon. It was like the sight and odor of cake to a schoolboy" (*MW* 86). In this passage, "primitive" carries an unambiguously positive connotation, suggesting firstness and originality, as in the term "primeval." Thoreau notes a dozen trees, shrubs, and forbs along the way, though there are few flowering plants in mid-September, and "Canada thistle, an introduced plant, was the prevailing weed all the way to the lake" (*MW* 87). He also notices "whole fields full of ferns, now rusty and withering, which in older countries are commonly confined to wet ground" (87). Already at the beginning of this second excursion, Thoreau displays a keener botanical eye and a more detailed appreciation than he shows in "Ktaadn."[7]

Thoreau's persistent strategy in "Chesuncook" is to portray his own botanizing as his principal activity. This portrayal occurs at least six times in the narrative. In its most telling uses, the strategy distinguishes Thoreau's botanical work from the moose hunting of his companion George Thatcher and their guide Joe Attean, the "murder of the moose" (*MW* 121), and the "tragical business" of dressing and skinning it (115). It highlights botany as a major element of the excursion, fully equal to Thatcher's goal of killing a moose. At the first carry, from the head of Moosehead Lake to the Penobscot River, Thoreau lets Thatcher walk ahead to hunt partridges while he follows, "looking at the plants." He describes the carry, along a railroad track, as "an interesting botanical locality for one coming from the South to commence with," for many of the plants are rare or completely absent in Massachusetts. He follows with a list of sixteen plants, many of which are indeed northern species: Labrador tea [*Ledum groenlandicum*], *Clintonia borealis, Linnaea borealis,* and others. Even familiar plants have "a peculiarly wild and primitive look," while the evergreen trees seem to crowd in toward the railroad tracks "to welcome us" (93–94). Interestingly, he mixes common and scientific names, and he does not italicize the scientific names, as if the distinction between nomenclatures should dissolve, so that *Clintonia borealis* (blue-bead lily) could become a commonly used term.

As the narrative continues, botanical moments multiply. Thoreau humorously notes, at the first campsite, that in the early morning he takes "a botanical account of stock of our domains, before the dew was off" (*MW* 105), and he adds that as they glide near the shore of the river, he frequently makes Attean turn aside so he can "pluck a plant" (106). Indeed, the "botanical account" dominates the narrative. Thoreau identifies many trees and shrubs along the dead water of the Penobscot (96–97, 98, 105–6), and his eyes are "all the while on the trees" (108), leading him to compare evergreens and deciduous hardwoods (108–9). The evergreens, he claims, are "Indian," while the birch, maple, elm, and beech are "Saxon and Norman." There is some tongue-in-cheek humor here, but the basic analogy to "civilized" and "savage," or "cultivated" and "primitive," is a predictable opposition, reductive and simplistic, to which Thoreau adds a racial component.

The murder of the moose intervenes. Thatcher shoots a cow, and, after some unsuccessful tracking and considerable luck, the three men find her. The travelers had completed a short portage on Pine Stream, and Thoreau describes himself as "absorbed in the plants, admiring the leaves of the aster macrophyllus, ten inches wide, and plucking the seeds of the great round-leaved orchis, when Joe exclaimed from the stream that he had killed a moose. He had found the cow moose lying dead" (*MW* 112–13). The juxtaposition between two spectacular northern plants and the dead animal is clearly ironic, but the irony remains understated until Thoreau describes their camp that night. The two hunters are still keen to shoot another moose in the moonlight, but Thoreau remains behind in camp, kindles a fire, and examines "by its light the botanical specimens which I had collected that afternoon, and wrote down some of the reflections which I have here expanded" (120).

In a letter of 23 January 1858 to James Russell Lowell at the *Atlantic Monthly,* Thoreau defined the subjects of "Chesuncook" as "the Moose, the Pine Tree, & the Indian."[8] Sitting by the campfire with his specimens and his expanding thoughts, Thoreau finds these three central ideas of the narrative. As he remembers the killing of the moose, he discovers that it has undermined the "innocence" of the excursion and "destroyed the pleasure of my adventure" (MW 119). He considers the ethics of hunting, the unequal balance of power, and the moose, "God's own horses, poor timid creatures that will run fast enough as soon as they smell you, though they *are* nine feet high" (119). And, he asks rhetorically, is hunting even a sport? Just as easily kill an ox with a gun. The equivalence of the moose to domestic animals leads to the very nature of killing itself. Questions abound: Why are men's motives for entering the wilderness so base and coarse? Why do explorers, lumbermen, white men, Indians—why do they all come as "hirelings" and insist on killing? Could they not find other "employments than these—employments perfectly sweet and innocent and ennobling?" (120). These questions lead Thoreau to sit by the campfire and determine the three major subjects of "Chesuncook": "What a coarse and imperfect use Indians and hunters make of nature! No wonder that their race is so soon exterminated. I already and for weeks afterward felt my nature the coarser for this part of my woodland experience, and was reminded that our life should be lived as tenderly and daintily as one would pluck a flower" (120). The idea of the "exterminated" race of Indians and hunters is undeniably racist, consigning Native Americans to the clichéd fate of an irretrievable oblivion. Thoreau may have Attean and Thatcher in mind more specifically, in which case the extermination is inclusively vengeful. Angry and scornful and coarsened himself, Thoreau uses the image of plucking the flower to bring himself back to a less coarse emotional state, even if, as he says, the feeling persists for weeks afterward. The passage strangely combines fastidiousness with a racist condemnation and eradication.

These raw reflections on the killing of the moose lead to the more extended treatment of "the Pine Tree." In two substantial paragraphs, Thoreau considers the "true use" of the pine, treating it as if it were botanical kin to the moose, the Indian, and the hunter. Analogy characteristically drives Thoreau's argument. He argues that the "truest" and "highest" use of any creature is not defined by its death: "Every creature is better alive than dead, men and moose and pine-trees, and he who understands it will rather preserve its life than destroy it" (MW 121). The pine may be converted into boards; the moose may be converted into meat; the Indian and the hunter may be converted into "a dead human carcass." But death is not the highest use of any creature. For this reason, the lumberman and the tanner do not make the "truest use" of the pine tree. "No! No! it is the poet," who loves the pines and "lets them stand" (121–22). The practical uses of the pine can be seen in the lumberyard, the carpenter's shop, the tannery, the lampblack factory, and the turpentine clearing, but they do not reveal the "highest use": "It is not their bones or hide or tallow that I love most. It is the living spirit of the tree, not its spirit of turpentine, with which I sympathize, and which heals my cuts. It is as immortal as I am, and perchance will go to as high a heaven, there to tower above me still" (122). The last sentence, crucial to Thoreau's argument, was famously deleted from the *Atlantic Monthly* publication and called down Thoreau's righteous wrath upon the head of editor James Russell Lowell.[9] The "highest use" argument assumes that every living creature has a "living spirit," perhaps immortal, and certainly as worthy of ethical consideration as a human spirit. In this case, Thoreau's vitalism underpins an ethic of preservation that is closely aligned to the ancient religious and ethical concept of ahimsa, or nonviolence, to do no harm to other living creatures.[10] It joins the traveler to the hunter and Indian, the moose and the pine, for all possess a "living spirit" that deserves to be preserved and protected.

The white pine maintains a steady presence in Thoreau's mind as the narrative of "Chesuncook" turns toward home. At Ansell Smith's clearing, he encounters twenty or thirty lumbermen and considers them as evidence of "a war against the pines, the only real Aroostook or Penobscot war" (*MW* 128). The clear purpose of mixing together these three wars is to make the "war against the pines" the most important and consequential one. But, in his hyperbolic rhetoric, Thoreau missteps. While the Aroostook War is a famously comic nonwar between Great Britain and the United States concerning the northern boundary of Maine, the "Penobscot war" is unclear, perhaps referring to a disastrous naval expedition during the Revolutionary War, or to the series of Abenaki wars in the seventeenth and eighteenth centuries.[11] In warning of the loggers' "war against the pines," Thoreau's humorous, undercutting allusion to human wars misses the mark. Worse, it may inadvertently pit the pine against the Indian, belittling the reality of the Penobscot inhabitants of the Maine woods.

Returning to Moosehead Lake and the North East carry, the "interesting botanical locality" that began the wilderness part of the excursion, Thoreau abandons his search for the Pine in order to focus on the Indian. He convinces Thatcher to camp with a group of three Indians on the carry, and for several pages he recounts his fascination with the men, their Abenaki language, its rhythms and sounds, Native place names, and stories of moose (*MW* 132–42). During much of the excursion, Thoreau's attitude toward the young guide, Joe Attean, tends toward the condescending.[12] At the North East carry, however, he revels in listening to the four Native men speak "this unaltered Indian language, which the white man cannot speak nor understand. We may suspect change and deterioration in almost every other particular, but the language which is so wholly unintelligible to us" (136). Despite his gross generalization about "change and deterioration," the language comes in living speech, along with laughter and jokes, and Thoreau feels he "stood, or rather lay, as near to the primitive man of America, that night, as any of its discoverers ever did" (137). Like so many words in Thoreau's vocabulary, "primitive" does double duty, suggesting both the firstness and the "savagery" of the Indigenous inhabitants. He picks up the subject once again when he and Thatcher arrive in Oldtown and pay a visit to the Indian Island, calling on Governor Neptune (145–51). While there, Thoreau notices "an abundance of weeds, indigenous and naturalized; more introduced weeds than useful vegetables, as the Indian is said to cultivate the vices rather than the virtues of the white man" (146). The analogy is not persuasive, either in its botany or in its ethnology; it borders on the foolish to suggest that all "useful vegetables" are those of the white man. In such a gratuitous judgment, moreover, both Pine and Indian seem to be slipping away from Thoreau's verbal grasp, as if he has missed an opportunity to become naturalized to the place.

"Chesuncook" concludes with a coda, like "Ktaadn," but in this instance the essay takes a pair of interesting turns in the argument. Thoreau begins with Alexander von Humboldt and his "chapter on the primitive forest," and this leads to the distinction between wild and tame forests, both in Concord and in Maine.[13] In retrospect, the traveler can find no truly wild forest in Concord; "civilized man . . . tames and cultivates to a certain extent the forest itself" (*MW* 151). By contrast, he claims, the Maine woods hardly show the hand of human beings at all, and, despite the misleading accounts of geographers, the North remains "an almost unbroken pine-forest" (153). As in most of "Ktaadn," Thoreau erases the continuous presence of Abenaki Indians in New England, focusing instead on the fact that the Maine woods are not so cultivated and tamed as the forests of Massachusetts. Despite ridding the landscape of Indigenous people in order to praise its wildness, Thoreau sees danger on the horizon for Maine and its continuous forests.

The evidence of loggers and lumbermen, sawmills and factories and model farms, suggests that Maine is going the way of Massachusetts (153–55). Ultimately, he understands where the gravest danger lies.

At this point in the argument, Thoreau takes a surprising turn. He argues that for permanent habitation he prefers the "smooth" landscape of Massachusetts to the rough and shaggy wilderness of Maine (*MW* 155). The unbroken wilderness appears too "simple, almost to barrenness." Better the complications of cultivation, and better now that "the poet's, commonly, is not a logger's path but a woodman's" (156). Just as the essay seems ready to conclude, however, Thoreau turns another time, by returning to the spirit of the pine: "But there are spirits of a yet more liberal culture, to whom no simplicity is barren. There are not only stately pines, but fragile flowers, like the orchises, commonly described as too delicate for cultivation, which derive their nutriment from the crudest mass of peat. These remind us, that, not only for strength, but also for beauty, the poet must, from time to time, travel the logger's path and the Indian's trail, to drink at some new and more bracing fountain of the muses, far in the recesses of the wilderness" (156). With the word "spirits," Thoreau returns to the critical moment in "Chesuncook": to the campfire and his reflections on the "true use" and the "living spirit of the tree." Though rough, the pines are stately, and now he recalls "fragile flowers, like the orchises," because they remind us, among other things, of the campfire reflection that "our life should be lived as tenderly and daintily as one would pluck a flower" (120). Life is more than killing. Furthermore, the orchises remind us that no path is simple, just as the Maine woods are not all stately pines. The poet must follow the logger and Indian because their trails lead to new sources of strength and beauty. That means to discover paths of transformation, even if the path to the "new and more bracing fountain of the muses" runs through a mug of spruce beer or the killing of a moose.

In the final paragraph of "Chesuncook," Thoreau turns one last time, against his own scorn and anger by the campfire. He calls for the establishment of "national preserves" of forests, so that "the bear and the panther, and some even of the hunter race" may be preserved as well (*MW* 156). The idea of national forests is prescient and visionary, though preserves and other forms of bounded land had been much in settlers' minds since colonial days.[14] And yet, at the same moment, we find in that last phrase a common form of nineteenth-century American racism, mingling the idea of national preserves with Native reservations and equating the Indigenous people with wild animals. Despite the racist cliché of the "hunter race" and the abhorrent hedging of "some even," the notion of the national preserves counters the idea of extermination that marred Thoreau's campfire reflections. "Chesuncook" ends on a fitful note. Thoreau is enlightened when compared to Charles Dickens's assertion that the "savage" should be "civilized off the face of the earth," a statement he quotes directly.[15] But, finally, he is most focused on "our forests" as a source of "inspiration and our own true recreation" (156). Thoreau's botany is a source of enlightenment, but any sense of transformation and "true recreation" in the botanical traveler flickers in the shadows.

A NEW LIGHT: WAYS OF KNOWING

In "The Allegash and East Branch," Thoreau follows the Indian's trail more closely than ever before. Joseph Polis is the center of this long, unfinished narrative, and among his many gifts to Thoreau is traditional environmental knowledge in the form of ethnobotany. As in "Chesuncook," the gift is by no means complete, nor does Thoreau accept it with complete reciprocity. Still, the relationship between Polis and Thoreau is the most important of the book, and it adds a singular depth to the botanical excursion as a narrative.

Unlike the young Joe Attean, Joseph Polis is an experienced guide and hunter, knowledgeable about the plants and animals of his home territory, and with mature political and religious interests that travel beyond the Maine woods or Indian Island and also return home. Polis is a traveler, moving between the overwhelmingly white world of Boston or Philadelphia and the Penobscot world of the Maine woods. He reaches back into a traditional culture he knows imperfectly but deeply, and he inhabits a world that is a complex layering of traditional and modern, Native and white.[16]

From the very beginning of "Allegash and East Branch," Polis acts as a subtle teacher. Early on the first day of the excursion, after Polis has given his white clients several names for birds, and the three travelers have paddled up the eastern shore of Moosehead Lake, the narrator defines the visitors' feeling of being "stage-passengers and tavern-lodgers" who become "suddenly naturalized there and presented with the freedom of the lakes and the woods" (MW 165). That is Polis's first gift, "presented" to his clients. Significantly, Thoreau uses the language of botany to express his feeling of arriving in the place, of becoming "suddenly naturalized." If we follow Robin Kimmerer's understanding of becoming "naturalized," however, we recall that it takes centuries to become naturalized to a place. This moment in the canoe echoes the effect of spruce beer in "Ktaadn," in which the youthful Thoreau feels as if he has "sucked at the very teats of Nature's pine-clad bosom" and quaffed "the sap of all Millinocket botany commingled" (*MW* 27). In both passages, the traveler favors sudden transformation over hard-won acceptance.

Education is prominent on the first day of the excursion, as is the enthusiasm of the student. Polis teaches the travelers the Penobscot names of Usnea lichen and the wood thrush, and Thoreau enthusiastically suggests that he would "like to go to school to him to learn his language, living on the Indian island the while; could not that be done?" (*MW* 168). "Oh, yer," says Polis, "good many do so." Polis is offering his student a nuanced lesson—be humble; you are not alone in your enthusiasm for traditional ways of knowing. When Thoreau asks how long such schooling would take, Polis tells him a week. The irony of that response goes unremarked. Instead, the narrator offers to "tell him all I knew, and he should tell me all he knew, to which he readily agreed" (168). The test of the narrative is to weigh these stores of knowledge and what they mean.

As narrator, Thoreau enjoys a definite advantage, since he determines whom and what he describes and how he characterizes the events of the narrative. But Polis exercises a weighty authority as the guide in "Allegash and East Branch." Without him, of course, Thoreau and his companion Edward Hoar would never make the ambitious canoe trip of this third excursion, and Polis shows them how indispensable he and his knowledge are on more than one occasion. The exchange of knowledge between Thoreau and Polis occurs, at times, in a more balanced way. At a rest stop on the shore of the Moosehead Lake, for instance, Thoreau names four plants, mixing common names with scientific names for genus and species. The last, *Cornus stolonifera*, is also named red osier, and Polis adds that the bark is "good to smoke, and [is] called *maquoxigill*, 'tobacco before white people came to this country, Indian tobacco'" (*MW* 170). The one plant takes on four names in three different languages, a rich layering of association and meaning, combining Western science and traditional environmental knowledge. But the possibility that *maquoxigill* may have deeper cultural meanings and functions does not appear in the narrative. The gift of traditional knowledge is presented in a limited form, determined by the narrator's limitations. Thoreau evokes a time "before white people came to this country," but he does nothing to develop Polis's offering of the idea. It is as if the white man's footsteps obliterate the Indian's trail.

As frustrating and fulfilling as such a moment may be, the excursion may be most persuasive when it does not make too much of the reciprocity of gifts. The relationship between Thoreau and Polis develops over the course of the narrative, and it involves a certain degree of schooling for both characters. When Polis tells the traditional story of Mount Kineo having been originally a cow moose, Thoreau criticizes both the story and the telling in condescending, racist language: "An Indian tells such a story as if he thought it deserved to have a good deal said about it, only he has not got it to say, and so he makes up for the deficiency by a drawling tone, long-windedness, and a dumb wonder which he hopes will be contagious" (MW 172). The exchange of knowledge is still significant here, but its ultimate failure lies with the narrator.[17]

Later, the travelers climb to the summit of Mount Kineo in the rain, sending Polis back to camp while they enjoy their adventure. They climb, take in the scenery of Moosehead Lake, and botanize thoroughly. Thoreau lists twelve plants, an exquisite collection of northern flowers, in a mixture of common and scientific nomenclature. As the pair descend, having "explored the wonders of the mountain," they meet Polis, "puffing and panting," and the narrator speculates on his shortness of breath from a vantage of cultural superiority: "I thought that superstition had something to do with his fatigue. Perhaps he believed that he was climbing over the back of a tremendous moose" (MW 177). In context, the botanical knowledge displayed by the narrator is contrasted with the "superstition" of the Indian. But the two have much in common, too. The condescending narrator calls the Indian's story one of "dumb wonder," while he congratulates himself and his companion for exploring the "wonders of the mountain." If both ways of knowing lead to wonder, why is one superior to another? In this short episode, the cultures and their contexts coexist uneasily, and the narrator exercises the judgmental authority of the white man, the botanical traveler. But who is to say which way of knowing is more effective? The story of Mount Kineo gives glimpses of both ways, and neither one emerges as an absolute form of knowledge.

The episode on Moosehead Lake suggests that Thoreau practices a narrative art of juxtaposition, as we have already seen in "Chesuncook" with the murder of the moose and the narrator's botanical reflections. In one of the most famous scenes of "Allegash and East Branch," Polis sings hymns for his guests before bedtime at the first campsite, and Thoreau praises his style and tone in unconsciously condescending terms. The songs show Polis's "simple faith," and his hymns are not "dark and savage, only mild and infantile" (MW 179). Directly after that statement of assumed cultural and racial superiority, we read the detailed scene of the phosphorescent wood. Thoreau awakes in the middle of the night and sees a piece of "dead moose-wood (*Acer striatum*)" glowing like a "ring of light. . . . a white and slumbering light, like a glow-worm's" (179). Thoreau's response is as significant as the phosphorescence. The material of the plant, in common and scientific naming, is transformed into light, and the narrator considers it a personal gift: "I little thought that there was such a light shining in the darkness of the wilderness for me" (180).[18]

The personal discovery gains more power as its significance widens and deepens in cultural terms. The next morning, Polis gives the white men the Penobscot name for the light—*Artoosoqu'*—and cites numerous other instances of light and sound observed by his "folks" in the woods (MW 180–81). This report strikes Thoreau, to his credit, as wholly credible and fundamentally true. All trace of racist condescension, whether conscious or not, disappears from the narrative. The botanist "let[s] science slide, and rejoice[s] in that light as if it had been a fellow-creature" (181). Like the pine tree in "Chesuncook," the phosphorescent wood makes "a believer" of the scientific traveler, who now knows "that the woods were not tenantless, but choke-full of honest spirits as good as myself any day,—not an empty chamber, in which chemistry was left to work alone, but an inhabited house,—and for a few moments I enjoyed fellowship with them." As momentary as Thoreau's conversion to an animistic, vitalist world may be, it signifies Polis's powerful authority as a teacher in the excursion narrative. He is at home in the "inhabited house" and recounts multiple experiences of his culture with the "honest spirits" of the woods. Thoreau does not place any credence in the Christian revelation, but "for a few moments" he completely accepts Polis as his spiritual guide: "I have much to learn of the Indian, nothing of the missionary. I am not sure but all that would tempt me to teach the Indian my religion would be his promise to teach me *his*. Long enough I had heard of irrelevant things; now at length I was glad to make acquaintance with the light that dwells in rotten wood. Where is all your knowledge gone to? It evaporates completely, for it has no depth" (181–82). In cultural terms, Thoreau places the exchange of knowledge completely in the Indian's hands. He abandons his earlier critique of Polis's stories, for in this case he is directly "making acquaintance" with the light. The question concerning "all your knowledge" seems directed at white Christian culture, but it is also directed at Thoreau himself. What does his botany amount to, after all, when faced with light shining out of rotten wood? The assumed superiority disappears with the knowledge. Both evaporate, for neither has any depth.

Later in the narrative, Thoreau extends the critique of Christianity to American progress and politics, echoing both "Chesuncook" and the phosphorescent wood episode: "The Anglo American can indeed cut down and grub up all this waving forest and make a stump speech and vote for Buchanan on its ruins, but he cannot converse with the spirit of the tree he fells—he cannot read the poetry and mythology which retire as he advances" (*MW* 229). Polis connects Thoreau to Penobscot ways of knowing "the spirit of the tree," and Thoreau understands those ways as grounded in a language of reciprocity and communication—a conversation and a reading.

Whatever Thoreau's personal state of enlightenment concerning Native Americans and traditional environmental knowledge, the fundamental point is that "Allegash and East Branch" engages in a conversation between cultures. By developing Polis as spokesperson for Penobscot ways of knowing and as the preeminent practical and spiritual guide in the Maine woods, the narrator brings Western science and traditional knowledge into a significant relationship. That relationship is only now becoming part of mainstream education in the United States, largely because of resistance from the disciplinary limits of Western science. Robin Wall Kimmerer makes this point clearly in a 2002 essay for *BioScience* in which she explains that "Western science is conducted in an academic culture in which nature is viewed strictly objectively," whereas traditional ecological knowledge (TEK) "is much more than the empirical information concerning ecological relationships." Traditional knowledge "is woven into and is inseparable from the social and spiritual context of the culture" and, as a result, "includes an ethic of reciprocal respect and obligations between humans and the nonhuman world. In indigenous science, nature is subject, not object. Such holistic ways of understanding the environment offer alternatives to the dominant consumptive values of Western societies."[19] For Thoreau, *Acer striatum* becomes *Artoosoqu'*, and two "fellow-creatures" engage in deeply significant conversation.

Polis teaches the travelers in several aspects of traditional knowledge. The primary mode is linguistic. By telling the travelers the names of streams and lakes, birds and other animals, and plants, Polis brings his culture to them without translation. As significant as the name itself, the explanation and application of the name often lead to conversation and create the opportunity for learning. For example, Polis teaches them the names for white spruce and black spruce, but just as important to him is the fact that black spruce roots should be used to sew canoes, not white spruce roots (*MW* 188). Later, Polis teaches them how to distinguish black spruce from white spruce, then shows how to find the black spruce roots and fashion string from them (203–4).

In another salient example, Polis teaches the travelers the name *Umbazookskus,* meaning Much Meadow River (*MW* 207). Thoreau finds the name especially apt for describing the habitat, and he follows this observation with a list of a half-dozen plants growing in the watery meadows—sedges, woolgrass, blue-flag, and the red osier. Thoreau recognizes "a great many clumps of a peculiar narrow-leaved willow (*Salix petiolaris*), which is common in our river meadows," and Polis tells him that the "musquash ate much of it" (207–8). Without commentary, Thoreau presents a layered conversation. In response to Polis's language teaching, Thoreau brings his local botanical knowledge to bear on the meadow river in Maine, recognizing the species of willow in its habitat; then Polis teaches him the relationship of the willow to the muskrat, or musquash. Just before this scene, Polis attempts to call in a musquash to kill and eat (206). The network of plant, animal, and human takes explicit shape, keyed by the Penobscot name for the river. Implicit in the scene, moreover, is the cooperative sharing of knowledge by Thoreau and Polis. Two ways of knowing come close to creating a third way.

The narrator remarks that Polis does not use the word "muskrat," as if somehow the name *musquash* has incantatory power. Perhaps in a similar way, in the closing pages of "Ktaadn," Thoreau mentions the musquash in Louis Neptune's canoe, using the image to wax poetic over his Ideal Indian, disappearing over the horizon (*MW* 78–79). But even though Thoreau is willing to idealize Polis in some passages, he does not do so to offer a conventional lament for the Lost Savage. At the farthest northern point of the excursion, for example, Thoreau enumerates twenty-two plants that he identifies on the shores of the Allagash, claiming that he names them because it is the group's farthest north (234–35). Polis then makes his own claim, that he is a "doctor" and can give a "medicinal use for every plant" Thoreau shows him. Without enumerating this second list, Thoreau asserts that Polis is "as good as his word." Furthermore, Polis has acquired his knowledge from an elder, and now he worries that the present generation has "lost a great deal" (235). Polis functions in the scene as a guardian of Penobscot culture, a student and teacher for his own people. As in the previous example, moreover, the episode summarizes an extensive, cooperative exchange of knowledge based in different cultures, languages, generations, and ways of knowing.

If my characterization of these botanical scenes appears idealistic, consider Thoreau's description of Polis immediately afterward. Clearly at his most assertive, Polis remarks that he could get back to Oldtown in three days, and he then explains the route he would take in winter, traveling over the lakes and streams on snowshoes. Thoreau is plainly awestruck: "What a wilderness walk for a man to take alone!" (*MW* 235). The idea of Polis's winter excursion leads the narrator to an outburst of Penobscot place names:

> It reminded me of Prometheus Bound. Here was travelling of the old heroic kind over the unaltered face of nature. From the Allegash, or Hemlock River, and Pongoquahem Lake, across great Apmoojenegamook, and leaving the Nerlumskeechticook Mountain on his left, he takes his way under the bear-haunted slopes of Souneunk and Ktaadn Mountains to Pamadumcook and Millinocket's inland seas, (where often gulls'-eggs may increase his store,) and so on to the forks at Nickatou, (*nia soseb* "we alone Joseph" seeing what our folks see,) ever pushing the boughs of the fir and spruce aside, with his load of furs, contending day and night, night and day, with the shaggy demon Vegetation, travelling through the mossy graveyard of trees. (236)

The ecstatic portrait of Polis as heroic traveler echoes earlier points in the excursion, including the phosphorescent wood episode—"seeing what our folks see." Just as important, the allusion to Penobscot *"nia soseb* 'we alone Joseph'"* recurs to the message map that Polis uses at his campsite on the Caucomgomoc River, which the travelers reached on Sunday, 26 July (199). Polis's use of the pictograph text fits within the Native tradition of *Awikhigan,* written messages that could travel over the waterways of Maine through the network of relations connecting Native American travelers and hunters.[20] At the Sunday campsite, Thoreau describes the drawing of a bear paddling a canoe, "which he said was the sign which had been used by his family always" (199). The inscription on the fir tree lists two previous stays, in July 1853 and July 1855, and suggests that Polis is identifying his home as Oldtown. As Jeffrey Cramer notes, Thoreau seems to misinterpret the Penobscot text, which is a recurring message rather than a transitory one, identifying Polis himself as "me Joseph" in the word "niasoseb."[21] At this early stage in the excursion, Thoreau does not fully grasp the forms of literacy practiced by Polis and other Penobscot hunters, nor does he see how Polis uses both Penobscot and English to identify himself: "*Niasoseb*

Polis *elioi*" and "Io. Polis." Instead, Thoreau focuses on the idea of "we alone Joseph," the image of the solitary wilderness traveler.

The portrait of Polis halts the excursion at this farthest northern place in order to forecast a return trip, a kind of odyssey. By specifically evoking *Prometheus Bound,* moreover, Thoreau recurs to his own translation of the Aeschylean play, published in the *Dial* in January 1843. In the play, Prometheus is much more than the trickster Titan who steals fire and bestows it on humankind; rather, he is a seer and a giver of knowledge, both past and future. The most important speeches of the play, in this regard, are the catalog of arts that Prometheus teaches humankind, including most aspects of what we call "civilization," and the prophecy of Io's "vexatious wandering," a narrative of "traveling of the old heroic kind," as Thoreau puts it.[22] In every part of the play, Prometheus is an eloquent figure of knowledge and of the defiant giving of knowledge. That is also why he is chained to the rocks and tortured, as Thoreau describes it on the summit ridge in "Ktaadn": "Such was Caucasus and the rock where Prometheus was bound. Aeschylus had no doubt visited such scenery as this" (*MW* 64). Thoreau thus figures Polis as a Promethean seer and teacher, contending in his odyssey with "the shaggy demon Vegetation." But at its most interesting in this moment, "The Allegash and East Branch" presents both the classical learning of a Massachusetts writer and the multiple signifying writings of a learned Penobscot man of Maine.

The journey home along the East Branch, and especially the rapids of Webster Stream, is full of adventure, and Thoreau never stops using the combination of place names, plants, and Polis's knowledge to develop the excursion narrative. But even though the "vexatious wandering" of Edward Hoar provides anxiety and excitement, the northernmost point of the journey is in fact the climax of the narrative. Thoreau reminds himself of this moment after the three travelers pass out of the rapids and cross dead-water lakes:

Ktaadn, near which we were to pass the next day, is said to mean "Highest Land." So much geography is there in their names. The Indian navigator naturally distinguishes by a name those parts of a stream where he has encountered quick water and falls, and again, the lakes and smooth water where he can rest his weary arms, since those are the most interesting and memorable parts to him. The very sight of the *Nerlumskeechticook,* or Dead-Water Mountains, a day's journey off over the forest, as we first saw them, must awaken in him pleasing memories. And not less interesting is it to the white traveler, when he is crossing a placid lake in these out-of-the-way woods, perhaps thinking that he is in some sense one of the earlier discoverers of it, to be reminded that it was thus well known and suitably named by Indian hunters perhaps a thousand years ago. (*MW* 270)

This passage offers myriad rippling memories. It recurs to the party's initial view of the mountain three days before and Polis's naming of it (225). More deeply, it recalls the ways in which geographical place names function in Penobscot culture, not only as descriptive terms but also as mnemonics, narrative placeholders, and spiritual realities. Next, the passage takes the reader of *The Maine Woods* back to "Ktaadn" and the significance of the mountain both for Native inhabitants and for white travelers. Finally, in the image of the "white traveler," it reminds us that the Penobscot hunters are the original knowers and namers of the landscape, and they remain the primary givers of knowledge. So much geography is there in their names. That geography embraces botany, and it shines a light in darkness with its gifts.

IN PLACE OF A CODA

"The Allegash and East Branch" ends abruptly, without the coda that summarized both "Ktaadn" and "Chesuncook." That is a sign of the incompleteness of the narrative, just as the datelines after the first day on Moosehead Lake suggest journal entries more than polished transitions. Paddling directly to Joseph Polis's home, the white travelers stay an hour before taking the train to Bangor: "This was the last I saw of Joe Polis. We took the last train and reached Bangor that night" (*MW* 297).

In place of a coda, we may first return to Thoreau's 23 January 1858 letter to James Russell Lowell, who had written to him asking for a narrative about the 1857 Maine excursion. Thoreau begins the letter with an apology for not replying promptly, and then he adds a deeper reason for the delay:

> I have been so busy surveying of late, that I have barely had time to "think" of your proposition, or ascertain what I might have for you.
>
> The more fatal objection to printing my most recent Maine-woods experience, is that my Indian guide, whose words & deeds I report most faithfully,—and they are the most interesting part of the story,—knows how to read, and takes a newspaper, so that I could not face him again.[23]

As we saw earlier, this letter leads Thoreau to offer Lowell the "Chesuncook" excursion and the three subjects of moose, pine tree, and Indian. More interesting, however, is Thoreau's "fatal objection" to have Polis as a reader of "The Allegash and East Branch." The revisions of the text occupied him even on his deathbed, and we do not know if he would have been willing for Polis to read the narrative as he left it. In the letter to Lowell, he seems to be worried that he would offend Polis, as if he had not given Polis the respect he is due. Polis as primary reader of "The Allegash and East Branch" presents us with yet another interesting role, since it would make him privy to the narrator's comments and faithful reports. Finally, in the statement "so that I could not face him again," Thoreau is clearly imagining another excursion with Polis as his teacher and guide.

Even though the narrative is unfinished, the complex characterization of Polis as guide, language and botany instructor, culture teacher, and heroic traveler is vivid.[24] In addition, Thoreau seems quite correct in seeing Polis's "words & deeds" as the most interesting part of the story. Moreover, Thoreau faithfully reports not only Polis's words and deeds but also his own, and these include the benighted as well as the enlightened responses to Polis as a contemporary representative of "the Indian" and of Penobscot culture. Polis gives Thoreau his most capacious, generous version of the Maine woods. To Thoreau's credit, he realizes what the gift entails, particularly in the juxtaposition of Western botany and Polis's traditional environmental knowledge.

In entries for the Journal of 4 and 5 March 1858, months after the end of the botanical excursion, Thoreau is still thinking about Polis and the place names of Maine. Studying Father Sebastian Rasle's dictionary of the Abenaki language, Thoreau contrasts the Penobscot language with Western scientific language and knowledge.[25] He uses the arbor vitae as an example: "It is not a *tree* of *life*. But there are twenty words for the tree and its different parts which the Indian gave, which are not in our botanies, which imply a more practical and vital science. He used it every day. He was well acquainted with its wood, and its bark, and its leaves" (*J* X:294). As we have seen, the fault with this general view of "the Indian" is that traditional environmental knowledge is characterized as vanished, held only in Father Rasle's incomplete dictionary, just as Thoreau sees "the Indian" as practically extinct. As he continues the reflection, however, he evokes Joseph Polis: "It was a new light when my guide gave me Indian names for things for which I had only scientific ones before. In proportion as I understood the language, I saw them from a new point of view" (X:295). Incompletely realized though it may be, "The Allegash and East Branch" is Thoreau's most successful botanical excursion, one that takes him farther toward "a new point of view" than he ever goes in other odysseys.

CHAPTER 2

Cape Cod and the Seven Excursions

Cape Cod, published in 1865, followed *The Maine Woods* in the sequence of Thoreau's posthumous book publications. If Emerson had a hand in compiling *Excursions* (1863), Sophia Thoreau was the main editor of *The Maine Woods* (1864), while she and Ellery Channing shared the work of editing *Cape Cod* (1865). Sophia and Channing probably made the decision to use the manuscript as Thoreau had left it, without adding the material of the 1857 Journal to the ten chapters as Thoreau had drafted them. That appears to have been a solid decision. After all, Thoreau had crafted the book by using the initial excursion of October 1849 as the narrative frame. Over time, he added material from subsequent excursions and his extensive historical research, fleshing out the original story to make it into a book-length narrative. Beginning as lectures and partially published in *Putnam's Monthly Magazine*, the manuscript had been subject to numerous authorial revisions over the years, so it closely represented Thoreau's intentions for the published work.[1] In many ways, it is more polished and finished than *The Maine Woods* or the miscellany *Excursions*. On the other hand, the book stops well short of *The Maine Woods* in investigating the role of Native Americans in America. The few allusions to the Nauset Indians come from early French and English explorers or from Christian missionaries. Thoreau uses these passages in "The Plains of Nauset" and "Provincetown" as excerpts to fill out the chapters, not as any considered account of Native American inhabitants. In that regard, it is significant that Thoreau's final excursion to Cape Cod took place in the summer of 1857, before he traveled to Maine and came under the powerful influence of Joseph Polis.

The number of excursions that Thoreau made to Cape Cod is subject to some debate. He mentions three visits in the first sentences of the book:

"I made a visit to Cape Cod in October, 1849, another the succeeding June, and another to Truro in July, 1855; the first and last time with a single companion, the second time alone" (*CC* 3). This statement is factually correct, as of 1855. However, Thoreau visited his friends Marston and Mary Watson in Plymouth and went to Clark's Island in July 1851, and he visited his friend Daniel Ricketson in New Bedford twice, in June 1856 and April 1857. These six trips to Cape Cod and its immediate vicinity yielded material that was incorporated in the finished manuscript and final published version of the book.

In June 1857, moreover, Thoreau returned to Plymouth to visit the Watsons and then proceeded to the Cape for a last excursion on his own, hiking over eighty miles in a week. He used "the cars" for one short stretch, but otherwise he went by foot from Manomet Point, just south of Plymouth, all the way to Provincetown, at the far fist of the Cape. For much of his walking, he used a map and compass to find his own way, often crossing the countryside to avoid settled paths and roads. Some of the excursion was a return to familiar territory, but much of it opened new landscapes to the traveler. This final, seventh excursion came at a critical juncture. Thoreau had been seriously ill for some five months, through the winter and spring of 1857, so the hike on Cape Cod was a test of his physical and mental endurance. Rain and fog accompanied Thoreau for much of the time. The excursion was also a kind of prelude for the longer, more ambitious excursion to Maine later that same summer. The Journal entries for the 1857 Cape Cod excursion flesh out the 1849 frame narrative and deepen the sense of the book as a multidimensional exploration. Together, the seven excursions show the posthumous book as a succession of experiences and narratives, while the addition of the 1857 journal narrative, an admitted liberty, enriches the development of Thoreau as a botanical explorer and open-minded traveler. Taken together, the multiple sources of *Cape Cod* render it a botanical excursion.

THE NARRATIVE FRAME OF 1849

Cape Cod opens famously with the chapter "The Shipwreck," set in October 1849. The narrator and his unnamed companion (Ellery Channing) are prevented from sailing from Boston to Provincetown by a fierce storm, which has wrecked a ship at Cohasset, a harbor town north of Plymouth, on the western shore of Cape Cod Bay. They decide to take the train from Boston to Cohasset, where they walk a mile to the beach with hundreds of other people. Some of the walkers are Irish immigrants, searching for their relatives among the dead and the living, attending the funeral later that day; some are locals, attracted by a major disaster in their community; two are travelers, Concord literary writers.

The account of the shipwreck of the Irish brig *St. John* is heart-rending in many ways, though Thoreau reports the scene in direct, understated language. He describes the eighteen to twenty large boxes in which bodies are being gathered on the beach: "I saw many marble feet and matted heads as the cloths were raised, and one livid, swollen and mangled body of a drowned girl—who probably had intended to go out to service in some American family—to which some rags still adhered, with a string, half concealed by the flesh, about its swollen neck" (*CC* 5). Another body is the "coiled-up wreck of a human hulk," but bloodless, with eyes "like the cabin windows of a stranded vessel, filled with sand" (6). The human wreckage is all-encompassing, chaotic, overwhelming.

In response, Thoreau turns away and walks along the shore to a cove. Here we see the first appearance of plants in the book:

> It appeared to us that there was enough rubbish to make the wreck of a large vessel in this cove alone, and that it would take many days to cart it off. It was several feet deep, and here and there was a bonnet or a jacket on it. In the very midst of the crowd about this wreck, there were men with carts busily collecting the sea-weed which the storm had cast up, and conveying it beyond the reach of the tide, though they were often obliged to separate fragments of clothing from it, and they might, at any moment, have found a human body under it. Drown who might, they did not forget that this weed was a valuable manure. This shipwreck had not produced a visible vibration in the fabric of society. (*CC* 7)

The ironies are stark in the juxtaposition of rubbish and seaweed, the carts being used to haul off each of these, and the possibility of human bodies being found among the fragments of clothing and kelp. The tone of the final sentence is surprisingly difficult to pinpoint. Thoreau does not seem to condemn society for not showing visible concern for the deaths of Irish immigrants. But the image of "the fabric of society" resonates with the bonnet, jacket, and other fragments of clothing, and the equivalence of human beings and weeds is difficult to read without a tone of mordant irony, even sarcasm. Later, and farther down shore, the travelers meet an old man and his son collecting the seaweed: "It was the wrecked weed that concerned him most, rock-weed, kelp, and sea-weed as he named them, which he carted to his barn-yard; and those bodies were to him but other weeds which the tide cast up, but which were of no use to him" (9). The wreckage is so deep and encompassing that it includes the seaweed, and the inclusion erases the distinction between rubbish, human bodies, and weeds. The language of comparison differs from the first set of metaphors, in which the narrator compares human bodies to wrecked ships. In the second passage, the plants remain plants, even though the old man regards the plants as more useful than the human bodies, which are "to him but other weeds," though "of no use to him."[2]

The old man's apparent lack of sympathy for the shipwrecked dead is striking, and it may be contagious. Like a good journalist, Thoreau questions survivors and bystanders, gathering as much information as he can about the shipwreck and its consequences. He thoroughly walks over the scene ashore, describing the wreckage and corpses in detail. But in summary the narrator finds the scene "not so impressive . . . as I might have expected." After all, he reasons, "if this was the law of Nature, why waste any time in awe or pity?" (CC 9). Perhaps the sheer number of dead, the horribly public mixing of rubbish, clothing, weeds, and bodies, overwhelms the narrator, who decides that "it is the individual and private that demands our sympathy" (9). Or perhaps Thoreau's strategy is to wring the reader's sympathy from the narrator's critical stance: by showing no "visible vibration in the fabric of society," his descriptions ensure that readers will feel vibrations of their own.

As the first chapter proceeds, Thoreau imagines that over time the inhabitants of the coast will be affected more deeply by the wreck and wreckage. He shifts into a story of many days later, when a beach-walker spies "something white" floating in the water. The object turns out to be the body of a woman, "which had risen in an upright position, whose white cap was blown back by the wind" (CC 10). This macabre vision leads to a more consoling one: "I saw that the beauty of the shore itself was wrecked for many a lonely walker there, until he could perceive, at last, how its beauty was enhanced by wrecks like this, and it acquired thus a rarer and sublimer beauty still" (10).

Thoreau ends "The Shipwreck" by shifting again, to a walking tour he took one summer (most likely the visit to the Watsons in July 1851) from Boston to the village of Hull, Nantasket Beach, and Cohasset Rocks (CC 11–14). It is unclear whether Thoreau counts this tour among his Cape Cod excursions, since all the places Thoreau mentions are north of Plymouth

and much closer to Boston than to Cape Cod. The walking tour marks a
tonal counterpoint to the opening story of the shipwreck; it creates a strong
portrait of fine weather, benign conditions, and pleasurable activities like
swimming, fishing, and horseback riding. Along the coast, inhabitants are
collecting "Irish moss," probably for use as a thickener in processing foods
(13). Thoreau does not pause to give the scientific name of the seaweed;
rather, the choice seems rhetorically effective. The colloquial name evokes
the wreck of the *St. John* and emphasizes the contrast between the opening
scene and the summer tour.

The first five chapters of *Cape Cod* focus on the October 1849 excursion,
with only an occasional aside mentioning the June 1850 trip. These are also
the chapters Thoreau first wrote as a course of three lectures, delivered in
several versions in January 1850 and again in the summer of 1850. Finally,
the first four chapters were published in three issues of *Putnam's Monthly
Magazine*.³ While Thoreau reports his observations of plants during the
excursion, and while the plants function tellingly in a few places in the text,
the first chapters are given over to general views of the landscape and its
present-day human inhabitants. Chapter 2, "Stage-Coach Views," recounts
the trip by train and stage from Cohasset to Orleans. A woman on the stage
informs Thoreau that the dominant cover on the roadside is called "poverty grass" because "it grew where nothing else would" (18). This initial
view, fleeting and superficial, is fitting for a stagecoach perspective. Thoreau
spends most of the chapter making fun of his guidebooks and his fellow
passengers, though as the stage rolls through mist and rain he returns to
plants in order to evoke treeless plains and "singular barren hills, stricken
with poverty grass" (19).

Thoreau's identification of poverty grass is intriguing, based as it is on a
fellow passenger's word. Later in the book, Thoreau gives the scientific name
Hudsonia tomentosa for poverty grass, and for the rest of his excursions he
assumes that two species of *Hudsonia* are poverty grass. Nowadays, professional botanists apply the common names "poverty grass" and "poverty oat
grass" to *Danthonia spicata*, a true member of *Poaceae*, the grass family. Could
Thoreau have mistaken the roadside plant, or might he have misapplied the
common name to the two *Hudsonia* species he encountered and described
repeatedly? He never mentions *Danthonia* in *Cape Cod*, though it occurs in
the same areas as *Hudsonia*.⁴ The questionable botany is significant in that
it suggests how strongly the common name resonates in the narrative. For
Thoreau, Cape Cod is a place stricken by poverty, human and botanical.

The next three chapters comprise one rainy day's hike from Orleans to Truro. The walk allows Thoreau to provide more detail in his descriptions of Cape Cod, but he also expands the narrative with extracts and digressions, both humorous and serious. In "The Plains of Nauset," for example, Thoreau mixes his outer impressions of the "apparently boundless plain" with a strong inner sense: "My spirits rose in proportion to the outward dreariness. The towns need to be ventilated. The gods would be pleased to see some pure flames from their altars. They are not to be appeased with cigar-smoke" (*CC* 32). He reflects tellingly, if briefly, on land ownership and the dispossession of Native Americans (33), only to move abruptly into prolonged satirical readings of the historical ministers of Eastham (33–43). At the end of the chapter, Thoreau offers a mock apology for the digressions: "There was no better way to make the reader realize how wide and peculiar that plain was, and how long it took to traverse it, than by inserting these long extracts in the midst of my narrative" (43).

If the narrative were to continue in this vein, readers might wish for an authentic dose of Laurence Sterne. Happily, "The Beach" and "The Wellfleet Oysterman" follow—two of the strongest chapters in the book. "The Beach" comprises an eight-mile walk from Eastham to Wellfleet. Thoreau slows down the narrative, dwelling on the way in which the landscape forms zones of vegetation and bare sand. The spine of both the landscape and the description is the sand bank that forms the walker's path and the "backbone of the Cape" (*CC* 47). Facing north toward Provincetown, Thoreau describes the wide beach on his right; the sand bank, a hundred feet high, on which he stands; a desert of shining sand to the left, eighty rods wide; and then a "region of vegetation—a succession of small hills and valleys covered with shrubbery, now glowing with the brightest imaginable autumnal tints" (48). The sand bank itself forms a bare plateau, a tableland that runs the length of the cape, some twenty-eight miles northwest: "In short, we were traversing a desert, with the view of an autumnal landscape of extraordinary brilliancy, a sort of Promised Land, on the one hand, and the ocean on the other" (48). The walk builds toward a climax of perception, unimpeded by other human characters. As on the climb of Katahdin, Thoreau encounters a "solitude . . . of the ocean and the desert combined," a seemingly necessary condition for the vision of a "Promised Land." Like the famous "Ktaadn" episode in the Burnt Lands, Thoreau realizes the vision most acutely in aftermath, as he meditates on the scene:

There I had got the Cape under me, as much as if I were riding it bare-backed. It was not as on the map, or seen from the stage-coach; but there I found it all out of doors, huge and real, Cape Cod! as it cannot be represented on a map, color it as you will; the thing itself, than which there is nothing more like it, no truer picture or account; which you cannot go further and see. I cannot remember what I thought before that it was. They commonly celebrate those beaches only which have a hotel on them, not those which have a humane house alone. But I wished to see that sea-shore where man's works are wrecks; to put up at the true Atlantic House, where the ocean is land-lord as well as sea-lord, and comes ashore without a wharf for the landing; where the crumbling land is the only invalid, or at best is but dry land, and that is all you can say of it. (50)

The "humane house" refers to the rescue shelters placed in slight hollows along the sand bank, and Thoreau characteristically prefers "the true Atlantic House," with all its authentic power. The "thing itself" is dangerous, just as the summit of Katahdin and the continuous forests of Maine expose the traveler to an ultimately dangerous reality. Most important, the stark contrast of ocean and autumnal plants creates the impression that the Cape is alive, a living being, one that cannot be represented by "a map, color it as you will."

Given this vision, we might expect that the autumnal tints would draw the writer's eye most strongly, but the vegetation that exerts imaginative power is surprising, recalling shipwrecks and odysseys. In a paragraph running for over two pages, Thoreau repeatedly evokes the fabulous character of seaweed—"this kelp, oar-weed, tangle, devil's-apron, sole-leather, or ribbon-weed" (CC 52). He creates a vision from some vessel's deck, evoking "this great brown apron, drifting half upright, and quite submerged through the green water, clasping a stone or a deep-sea mussel in its unearthly fingers." The description is precisely imagined: the movement, changing form, drift and submergence of the plants, a forest; the root-like holdfasts of the kelp grabbing onto some stone or mussel at the seabed.

The sentences move into multiple sightings of the "cable-like weed," buoying up some treasure that turns out to be "ridiculous bits of wood or weed" (CC 52). Thoreau keeps his descriptive prose moving like the kelp, without apparent logic or direction but with an undercurrent of irony. If the seaweed seems to offer treasure, the treasure is revealed to be ridiculous.

And yet, despite the narrator's best efforts, the seaweed becomes ultimately "a singularly marine and fabulous product, a fit invention for Neptune to adorn his car with, or a freak of Proteus. All that is told of the sea has a fabulous sound, to an inhabitant of the land, and all its products have a certain fabulous quality, as if they belonged to another planet, from sea-weed to a sailor's yarn, or a fish story" (52). This image of the floating cables of kelp reminds of the phosphorescent wood in some ways, and in both cases the botanical object becomes something more than an object. Here, it is "fabulous," and it even sounds fabulous. Then it stands for all the products of the sea, and they all have a fabulous quality, an otherworldly quality, "as if they belonged to another planet." But, like the phosphorescent wood or the pine tree, the kelp is itself—until it becomes something else. The seaweed becomes "a sailor's yarn, or a fish story." The plant becomes a tale or legend, even a myth, a "fit invention for Neptune," or "a freak of Proteus."

It is certainly worth highlighting Thoreau's botany in this lengthy passage. The observation and description are precise and sensory, grounding and focusing the move into the fabulous. The whole passage is marked by vital, dynamic action and motion, evoking a powerful force that the narrator allies with the classical gods of the sea. Indeed, the protean quality of the seaweed becomes a synecdoche for the sea itself and all its products, as if the entire planet—and other planets, too—finds its model in the vitality of the plant. This is description evoking *phusis,* the "unearthly" power of the earth. Even though "The Beach" predates the botanical turn of 1850–51, it shows clear indications of the direction Thoreau's thinking and writing will take.

Thoreau gets physically close to the kelp, describing it in detail and sitting down to "whittle up a fathom or two of it, that I might become more intimately acquainted with it, see how it cut, and if it were hollow all the way through" (*CC* 53). Intimacy only increases the fabulous qualities of the plant, leading the writer to fables and tales. As if he needed support for his fascination, Thoreau quotes at length Henry Wadsworth Longfellow's poem "Seaweed." He gives us the first four stanzas, in which the poet evokes the plant itself, drifting all over the globe on the currents of the "restless main." Then Thoreau alters the final stanza and complicates Longfellow's basic analogy. For the poet as for the Cape Cod traveler, the seaweed floats in the ocean just as fragments of songs float in a poet's imagination. Then Thoreau revises the final stanza, distinguishing his seaweed from Longfellow's: "*These* weeds were the symbols of those grotesque and fabulous thoughts which have not yet got into the sheltered coves of literature"

(54). Unlike the comfortable repose of "household words" and "sheltered coves" so loved by Longfellow, Thoreau seeks the "not yet"—the creative and dangerous process of drifting, shifting, restless actions and "grotesque and fabulous" thoughts.

This long paragraph, so dynamic and visionary, moves in a drift around the central image of the seaweed. It bears emphasizing that *these* weeds assume the principal role of thinking, the plant-thinking that is the real "not yet" of Thoreau's excursion. The seaweed becomes the dynamic embodiment of the sea, and both appear "as if they belonged to another planet, from sea-weed to a sailor's yarn, or a fish story. In this element the animal and vegetable kingdoms meet and are strangely mingled" (52). Thoreau's strange minglings create the most dramatic instance of plant-thinking in *Cape Cod:* it is active but not goal-driven, and certainly not driven toward safe harbors in conventional writing.

Cape Cod is not always so dramatic, so given over to the exploration of Thoreau's botany, nor so suggestive of passages in the Journal of 1850 and 1851. The character sketch of "The Wellfleet Oysterman," chapter 5, mingles the 1849 narrative with elements from the June 1850 excursion and the June 1856 visit to Daniel Ricketson in New Bedford. Thoreau wisely lets the oysterman, eighty-eight-year-old John Newcomb, do most of the talking and storytelling. The result is a humorous, heart-warming portrait of a relic from the Revolutionary War. The oysterman discourses at length and breadth about his life and experiences, which run from Bunker Hill and General Washington on horseback to beach peas, apples, and how to poison a cat—or a traveler—with a sea clam.

Thoreau frankly enjoys the old man's palaver and pronounces him "the merriest old man that we had ever seen, and one of the best preserved. His style of conversation was coarse and plain enough to have suited Rabelais. He would have made a good Panurge" (CC 71). After liberally spraying the hearth and breakfast dishes with tobacco juice, Newcomb walks the travelers outside on the morning of 12 October. Thoreau has asked the names of so many plants that Newcomb makes his guest name the specimens, wild and cultivated, in his garden. The botanical traveler names three garden vegetables, grown from seeds gathered from the wreck of the *Franklin* the previous year, and one wonders whether the oysterman really does not know the identity of cabbage, broccoli, and parsley. More likely, this is an old man's test. Thoreau also names eight other plants by their common names, besides the "common garden vegetables." These eight are all cultivated plants that have escaped and been introduced into disturbed habitats. As such, they seem to border between the wild and cultivated. Some are common, while others are not so well known, especially not by their common names. Even a seasoned botanist may not immediately recognize "Elecampane" as *Inula helenium* or know that it is introduced from Europe, cultivated, escaped, and sparingly naturalized. Once again, this is a test, but of whom? If Thoreau passes the test by naming the eight plants, we modern readers must rely on field guides and websites to help us pass muster.[5] And if this episode in fact took place in October 1849, Thoreau displays more botanical knowledge than we would expect, well before the botanical turn of 1850–51. The scene in Newcomb's garden may show signs of later revisions.

SUCCESSION AND THE LATER EXCURSIONS

The second half of *Cape Cod* differs from the first in at least two related ways. The five chapters feature more interpolated material, much of it from July 1855 and after. In addition, the chapters feature plants and botanical ideas much more prominently than the first five chapters. The result is that *Cape Cod* becomes a botanical excursion in much the same way that *The Maine Woods* does: by a revisionary process of succession and accretion. Most important, plants provide both the material and the process for the revised strategies of the book. In the opening of chapter 6, "The Beach Again," Thoreau describes the northern bayberry shrub at length and recounts making tallow from the berries one April (*CC* 80–81). This extended paragraph comes from the Journal entries of 7–8 April 1857, during Thoreau's second visit to Daniel Ricketson in New Bedford (*J* IX:320–21). The narrator had earlier noted the bayberry, along with beachgrass and other common plants, in "The Beach" (*CC* 44, 48), but here it becomes an image of a particularly useful plant. Tallow, after all, is used to waterproof hiking boots, whether on the Cape, in Maine, or at home in Massachusetts. The 1857 addition to the narrative reveals Thoreau's successive revisionary strategies at work, strategies that often depend on his growing botanical knowledge in the 1850s.

Not all examples of succession are botanical. In one remarkable passage of "The Beach Again," Thoreau describes beachcombing near Truro the day that he and Channing depart from Wellfleet. Objects on the beach appear larger and more imposing than they turn out to be, and Thoreau cites a "mass of tow-cloth" from the *Franklin,* which had wrecked months before their October visit (*CC* 83–84). He adds another mirage on a beach, one that turns out to be a heap of rags from some other unnamed shipwreck (84). Finally, he details his search for "the relics of a human body, mangled by sharks." This last passage stems at least partly from the Journal of 1850, and it appears to recount aspects of Thoreau's July 1850 trip to Fire Island, in search of the remains of Margaret Fuller and her family (*PJ* 3:94–95).[6] The salient aspect of the description is the way in which the "insignificant sliver" of humanity looms large on the beach, and how, once the narrator reaches the body, the remains grow "more and more imposing" (*CC* 84). As in "The Shipwreck," the narrator expresses his distance from the scene of death, but here he upbraids his "sniveling sympathies" in the face of the corpse: "That dead body had taken possession of the shore, and reigned over it as no living one could, in the name of a certain majesty that belonged to it" (84–85). The 1850 episode lends considerable dignity to the act of beachcombing, partly because Thoreau keeps both the "relics" and the source of the story anonymous. It is as if the travelers might stumble across such a dignifying relic in the present moment.

In general, natural history reigns over the chapter "The Beach Again," and it seems to be a direct result of the two-week stay at the Highland Lighthouse in July 1855. If one does not read carefully, the narrative makes it seem as if, on one day's walk, Channing and Thoreau encounter dozens of species of mollusks, crustaceans, echinoderms (*Radiata,* in Thoreau's nineteenth-century terminology), plants, birds, and fish. But Thoreau gives subtle temporal signals that the multitude of sightings come from July 1855. This, of course, makes perfect sense. During the two weeks of boarding with James Small at Highland Lighthouse, Channing and Thoreau were able to take repeated walks in the area and amass dozens of observations. The Journal for the two weeks fills some twelve pages with identifications, especially of birds and plants. The identifications of plants characteristically feature the Latin binomial as well as a common name, whether the plant is blossoming or fruiting, and any physical characteristics that aid in the identification. These passages in the Journal also express Thoreau's confidence in his botanical knowledge. When Thoreau queries his identification of a plant, it is in technical terms and often simply a question of species, not family or genus. As a traveling collector, he exhibits considerable expertise. Indeed, compared to the Journal entries, the list of plants in "The Beach Again" is merely an interesting compendium, the tip of a botanical iceberg.

Even if we see the revisions of *Cape Cod* as moving the book in the direction of the botanical excursion, portraying the traveler as an accomplished botanist and naturalist, it is quite clear that Thoreau is as interested in human culture as he is in plants. He is as keen to evoke the aesthetic and spiritual experience of beachcombing as he is to list the species of plants and animals he finds along the shore. As he says toward the end of "The Beach Again," "Nothing memorable was ever discovered in a prosaic mood" (*CC* 95), and he cites both Humboldt and Darwin as poetic explorers and discoverers (95–96). In the next chapter, "Across the Cape," he emulates his two heroes in finding beauty among the dwarf trees and resilient plants of the interior. Plants like bearberry and pitch pine evoke alpine tundra and krummholz, but Thoreau also notes that the following summer (1850) he saw "sylvan retreats . . . where small rustling groves of oaks and locusts and whispering pines, on perfectly level ground, made a little paradise" (101). To be accurate, Thoreau admits that the trees are "scraggy shrubbery," and that bayberry and wild roses form patchy thickets. But he also notes that these mixed patches, "in the midst of the sand, displayed such a profusion of blossoms, mingled with the aroma of the bayberry, that no Italian or other artificial rose-garden could equal them. They were perfectly Elysian, and realized my idea of an oasis in the desert" (102). In addition to mixing excursions, Thoreau's descriptions merge realistic observation and transformative description. They also connect the chapter to "The Beach" and the description of the autumnal plants as a "Promised Land" (48).

The succession of excursions allows Thoreau to offer moderating views of the Cape, informed by his more extensive and intimate experiences after 1849. So, for instance, while traversing the uplands, he praises poverty grass for its summer growth, in a passage that likely comes from the 1855 excursion:

> In summer, if the poverty-grass grows at the head of a Hollow looking toward the sea, in a bleak position where the wind rushes up, the northern or exposed half of the tuft is sometimes all black and dead like an oven-broom, while the opposite half is yellow with blossoms, the whole hillside thus presenting a remarkable contrast when seen from the poverty-stricken and the flourishing side. This plant, which in many places would be esteemed an ornament, is here despised by many on account of its being associated with barrenness. It might well be adopted for the Barnstable coat-of-arms, in a field *sableux*. I should be proud of it. Here and there were tracts of Beach-grass mingled with the Sea-side Golden-rod and Beach-pea, which reminded us still more forcibly of the ocean. (*CC* 106)

Lightly humorous, this passage insists that a plant is to be judged, if at all, when seen from every perspective. Thoreau seems to esteem poverty grass as a plant to be proud of, for it manages to flourish in the sandiest of soils. But just as the identity of the plant is open to question, so is the viewpoint of whether the plant is flourishing or poverty-stricken. The mingling of perspectives works well with the final sentence of the paragraph, which notes the mingling of three of the most common seaside plants, the same three Thoreau had included in his list of plants in "The Beach Again": beachgrass, seaside goldenrod, and beach pea (87). In a later instance of transforming perspective, Thoreau sees a homely pitch pine plantation as an important and successful experiment: "It seemed a nobler kind of grain to raise than corn even" (108). A consistent pattern in Thoreau's descriptions of common plants is his effort to ennoble them in his reader's eyes, and that pattern emerges most clearly from the succession of excursions across seemingly barren landscapes.

"The Highland Light" focuses on the part of Cape Cod that Thoreau knows most deeply, for he made four separate trips to the place, including the two-week stay at the lighthouse in July 1855. In the chapter devoted to the lighthouse itself, he includes lengthy descriptions of the Clay Pounds, an immense bed of clay stretching across the cape, from the lighthouse to Cape Cod Bay. Thoreau finds unexpectedly fertile ground and abundant plants (he lists over twenty species) in his walks across the Clay Pounds. The plants are "dwarfish" but fruitful, and this observation leads him to discuss the dispersion of seeds when he sees beets and turnips coming up in the seaweed (*CC* 130). He speculates that shipwrecked cargoes could disperse seeds on desolate islands, and that the plants would become "naturalized and perhaps drive out the native plants at last, and so fit the land for the habitation of man. . . . Or winds and currents might effect the same without the intervention of man" (131). Indeed, he thinks, perhaps some of these common succulent plants on Cape Cod are themselves naturalized shipwrecks! In passages like this one, we see Thoreau writing in the late mode of *The Dispersion of Seeds* and "Succession of Forest Trees," using his botanical observations as a means of thinking through larger ecological patterns and processes. Like his thinking in those late essays, he persistently relates plants to human beings while also entertaining ideas that do not require human intervention in natural processes.

Following his speculations concerning common plants on the seashore, Thoreau lists seven more plants he finds around the lighthouse in the

summer of 1855. Several of these he observes and describes in the Journal
of 7–8 July 1855 (*J* VII:434–36). One, broom crowberry (*Corema Conradii*), he
notes is supposed to be found only in Plymouth, but he finds pretty mounds
of the plant again in Provincetown. Then, "prettiest of all," he finds the
"scarlet pimpernel, or poor-man's weather-glass (*Anagallis arvensis*)" (131). He
adds two more plants sent to him from Yarmouth on 7 September, without
a specific year (131). These are points not originally made in the Journal but
added in drafting the book. Thus, in one short passage, Thoreau's botany
encompasses theoretical speculation, geographical distribution, aesthetic
appreciation, and a community of botanists with whom he corresponds.

The layering effects of multiple excursions and successive drafting are
clear in chapter 9, "The Sea and the Desert." The frame narrative is a two-
day hike on 13–14 October 1849, from the Highland Lighthouse along the
coast and across the sandy spine to Provincetown for the night, then northeast, back to the coast, before the return to Provincetown for a second night.
As in previous chapters, however, this crisscrossing journey gathers depth
because of the interpolated material from other excursions. The chapter
opens with an appreciation of the "resounding sea" and the mackerel fleet;
then Thoreau interpolates the story of his trip to Clark's Island aboard a
mackerel schooner (*CC* 141–45). The Clark's Island story comes from Thoreau's walking excursion to Plymouth by way of Cohasset and Duxbury,
taken in July 1851 (*PJ* 3:331–50).

Toward sunset on 13 October, Channing and Thoreau cross from the
seashore to the eastern edge of Provincetown, making their way across "the
Desert" (*CC* 152). The view from a barren sandhill leads Thoreau to describe
the autumnal landscape as a remarkably beautiful tapestry:

> There was the incredibly bright red of the Huckleberry, and the reddish
> brown of the Bayberry, mingled with the bright and living green of small
> Pitch-Pines, and also the duller green of the Bayberry, Boxberry, and Plum,
> the yellowish green of the Shrub Oaks, and the various golden and yellow
> and fawn colored tints of the Birch and Maple and Aspen,—each making its
> own figure, and, in the midst, the few yellow sand-slides on the sides of the
> hills looked like the white floor seen through rents in the rug. Coming from
> the country as I did, and many autumnal woods as I had seen, this was per-
> haps the most novel and remarkable sight that I saw on the Cape. (152–53)

This passage accumulates its power from the vivid descriptions of colors and
adds to that power by naming the plants with capital letters, marking the

species as individual types or characters in the scene, "each making its own figure." The cumulative effect recalls earlier descriptions of autumnal tints from "The Beach" and "Across the Cape," and most of the plants listed here are familiar to the reader from multiple encounters in earlier chapters. The autumn colors gather a further, tongue-in-cheek power in the final chapter of *Cape Cod*, in which Thoreau "*suspect[s]* that fall is the best season" to visit the Cape because of the clear air and "autumnal tints, such as, methinks, only a Cape Cod landscape ever wears" (214). Such a judgment, even as a tossed-off remark, comes only after many excursions over many years.

On the second day of the "Sea and the Desert" chapter, 14 October, Thoreau focuses his attention on beachgrass, although his descriptions of the day include passages from the June 1850 and July 1855 excursions describing other plants. The beachgrass is an important dune builder, and Thoreau notes that it has been planted by the government to create ridges and hills to combat the force of the winds (*CC* 160–64). The descriptions reach back to previous centuries of planting, to early narratives like Timothy Dwight's *Travels in New-England and New-York* and John Gerard's *The Herball* (162–63). They clearly include both native (*Ammophila breviligulata*) and introduced species of beachgrass (*Ammophila arenaria*), since some seeds are imported from Holland (164). Indeed, beachgrass is ubiquitous in the narrative of *Cape Cod*, appearing in most of the chapters and dominating the beaches and sandhills. It is most like beach heather (*Hudsonia tomentosa*) or poverty grass (*Danthonia spicata*), so common as to be unappreciated or even lamented, but for Thoreau it is an absolute necessity: "Thus Cape Cod is anchored to the heavens, as it were, by a myriad little cables of beach-grass, and, if they should fail, would become a total wreck, and erelong go to the bottom" (164).

THE OMITTED SEVENTH EXCURSION

The second half of *Cape Cod* shows clear elements of Thoreau's botany and the ways in which it affects the revisions of the book, adding layers of experience and meaning to the traveler's botanical excursion. But the six excursions that lurk within the pages of *Cape Cod* are not the only botanical excursions Thoreau took to the Cape. Omitted from the final edition of the book is the seventh excursion, the most coherent and developed botanical excursion Thoreau undertook before his last trip to Maine.

Thoreau's seventh excursion to Cape Cod begins on 12 June 1857, with a visit to the "Natural History Rooms" at Harvard University. The Journal notes observations on birds' eggs and an error by François André Michaux regarding the flowering of the red maple: "He also says that 'the red-flowering maple [*Acer rubrum*] is the earliest tree whose bloom announces the return of Spring.' This is a mistake, the white maple being much earlier" (*J* IX:414).[7] By that afternoon, Thoreau is in Plymouth, staying with his friends Marston and Mary Watson.[8] In his page of notes on the Watsons' extensive gardens and orchards, Thoreau lists over fifteen species of trees that he examines and from which he collects herbarium specimens. He adds another group of familiar flowering plants like bayberry, beach pea, and *Senecio vulgaris,* or common groundsel, an Old World native that he calls "a common weed" (IX:414–15; see *G&C* 559).

Over the next two days, Thoreau visits the beach and accompanies the Watsons to Clark's Island, mixing botanical observations with humorous stories (*J* IX:415–20). He notes on June 13 that he sees "children going a-flagging and returning with large bundles, for the sake of the inmost tender blade. They go miles for them here" (IX:415). The cattail flags also appear in *Cape Cod* and in the July 1855 Journal. In the entry for 7 July 1855, Thoreau writes that his host James Small "says there are two kinds of cat-tail there, one the barrel for coopers, the other shorter for chairs; he used to gather them" (*J* VII:435). This fact appears in the chapter "Across the Cape" as a one-sentence paragraph: "At the Pond Village we saw a pond three eighths of a mile long densely filled with cat-tail flags, seven feet high,—enough for all the coopers in New England" (*CC* 111). In the 1857 excursion, it is unclear whether the children are collecting cattail flags for barrels or chair bottoms, but clearly Thoreau is observing them in an activity he readily recognizes as familiar. The humorous stories include Daniel Webster's secret sighting of a sea monster and a certain General Winslow's adventurous ride home after a drunken party at the Plymouth Lighthouse on Gurnet Point (*J* IX:415–17). Following visits to friends and neighbors, the Watsons take Thoreau by carriage to Manomet Point on the afternoon of 15 June: "There I shouldered my pack and took leave of my friends,—who thought it was a dreary place to leave me,—and my journey along the shore was begun" (IX:420).

For the next eight days, Thoreau walks some eighty miles to Provincetown, with only a short train ride from Scusset village to Sandwich on 16 June. For most of the trip, Thoreau stays away from the main roads and cart-paths. On the first day, for instance, he notes that he has walked "six or seven miles from Manomet through a singularly out-of-the-way region, of which you wonder if it is ever represented in the legislature" (*J* IX:422). He often asks for directions, but just as often he disregards recommendations and prefers finding his own route with map and compass:

> I have found the compass and chart safer guides than the inhabitants, though the latter universally abuse the maps. I do not love to go through a village street any more than a cottage yard. I feel that I am there only by sufferance; but I love to go by the villages by my own road, seeing them from one side, as I do theoretically. When I go through a village, my legs ache at the prospect of the hard graveled walk. I go by the tavern with its porch full of gazers, and meet a miss taking a walk or the doctor in his sulky, and for half an hour I feel as strange as if I were in a town in China; but soon I am at home in the wide world again, and my feet rebound from the yielding turf. (IX:428)

This passage, characteristically wry, recalls a similar moment in "The Plains of Nauset," in which the traveler feels "mean and disgraced" in the towns because of the "savage and filthy habits" of smoking and drinking he encounters there. Only when he escapes the town do his spirits rise, "in proportion to the outward dreariness" (*CC* 32). Only in expanses marked by solitude does he have visions of dynamic nature (48–50). Thoreau insists on his independent perspective as a traveler, for in seeing the world "from one side" he finds his home.

The sense of familiarity, of being "at home in the wide world again," recurs in the pages of the 1857 Journal. As he walks, Thoreau thinks about how often, "summer and winter and far inland, I call to mind that peculiar prolonged cry of the upland plover on the bare heaths of Truro in July" (*J* IX:427). This brief note alludes to a much longer Journal entry for 12 July 1855, in which the description of the upland plover's "quivering note" leads Thoreau to quote the *Manual for Young Sportsmen* by Frank Forester (Henry William Herbert) (*J* VII:439–40). The 1855 entry is the source for Thoreau's description of the upland plover's cry in "Across the Cape" (*CC* 103), and both descriptions echo strongly in the 1857 Journal: "Though you cannot guess how far the bird may be, as if it were the characteristic sound of the Cape" (*J* IX:427). As he walks, the traveler remembers the cry of the upland plover, and he remembers how often he has remembered it in other times and places. If that layered memory creates a feeling of familiarity, it may deliver a sense of hauntedness as well. Both are characteristic of the Cape and of the traveler's perceptions.

In other passages, observations and memories create the simple pleasure of recognition. Approaching Scusset village from the shore, for instance, Thoreau sees an old friend: "Handsome patches of *Hudsonia tomentosa* (not yet had seen the *ericoides*), its fine bright-yellow flowers open chiefly about the edges of the hemispherical mounds" (*J* IX:427). Thoreau notes both species of *Hudsonia* in "Across the Cape," and near the beach he describes the yellow blossoms and "hemispherical tufts or islets," praising the beauty of the plant (*CC* 105–6). In the 1857 excursion, he is intent on distinguishing the two species of *Hudsonia,* their characteristic habitat, and how prolific they are. So, for example, he registers a kind of botanical excitement in the sentences, "The pine and oak woods were quite extensive, but the trees small. See the *Hudsonia ericoides,* with a *peduncle*" (*J* IX:429). This sighting accords with the observations of professional botanists: *H. tomentosa* is locally abundant on coastal dunes, while *H. ericoides* is locally abundant in pine barrens; moreover, *H. ericoides* features pedicels (or peduncles) of 5–15 millimeters (*G&C* 155). But what professional botanist would exclaim, "With a *peduncle!*"

As the 1857 excursion progresses, Thoreau builds on accumulating recognitions and reflections. Crossing the elbow of the Cape, from Harwich to Brewster, he once again reflects on his use of chart and compass, his connection with different landscapes, and how these create a larger idea of direction. A woman tells him that if he wants to get to Brewster, "I had come a good deal out of my way." Thoreau is incensed: "They take me for a roadster, and do not know where *my* way is" (*J* IX:431). Then he reflects more deeply on what his way might be: "I go along the settled road, where the houses are interspersed with woods, in an unaccountably desponding mood, but when I come out upon a bare and solitary heath am at once exhilarated. This is a common experience in my travelling. I plod along, thinking what a miserable world this is and what miserable fellows we that inhabit it, wondering what it is tempts men to live in it; but anon I leave the towns behind and am lost in some boundless heath, and life becomes gradually more tolerable, if not even glorious" (IX:431–32). The "boundless heath," on Cape Cod at least, is marked by mounds of *Hudsonia,* low shrubs like bayberry and blueberry, and ground cover like bearberry and *Cladonia* lichens. For Thoreau in the summer of 1857, this is a homecoming, and he comes home to his sense of boundless, independent travel. As he crosses the Cape, passing cranberry meadows, ponds, and one "noble lake, some two miles long," the traveler makes his way across solitary heaths and through extensive, low woods of pine and oak. He arrives at a secluded cottage, knocks all around the house, and finally meets "a woman with a child in her arms there ready to answer my questions." In this encounter, Thoreau does not contradict the young mother or praise his own independent way-finding. Rather, with great simplicity, he remarks, "I found that I had not come out of my way" (IX:435).

The following day, 18 June, Thoreau walks from Brewster to the High-land Lighthouse in Truro, spending three nights with James Small, his land-lord from July 1855. It is "a mizzling and rainy day with thick driving fog; a drizzling rain, or 'drisk,' as one called it" (J IX:437). By midday, Thoreau is soaked, stops briefly for shelter in a "Humane house" in Newcomb's Hollow, and arrives at a house near John Newcomb's, only to find that the Wellfleet Oysterman has died the previous winter (IX:438). The next day goes better. The pages of the Journal come alive with extended discussions of the sea and its wave actions ("a contra-dance to the shore," IX:443); pov-erty grass, in both species; cattail flags, both barrel and chair (IX:443–44). His voluble host James Small takes over the foggy, rainy days with his many visitors and conversations. The skies clear around noon of the 21st, and Thoreau sets out for Provincetown, crossing the Cape toward the Bay. Once he strikes the barren, solitary plain, his spirits lift, just as they did on the way to Brewster: "That solitude was sweet to me as a flower. I sat down on the boundless level and enjoyed the solitude, drank it in, the medicine for which I had pined, worth more than the bear-berry so common on the Cape" (IX:450). This is a fine moment of plant-thinking, combining the first simile with the medicinal pun and the botanical sense of value we know well from "Chesuncook" and "The Allegash and East Branch."

Approaching Provincetown and the end of the 1857 botanical excursion, Thoreau turns out of his way to climb the only hill, Mt. Ararat, for the view. It is the final description of landscape on the trip and a suitable ending to the narrative:

It exhibited a remarkable landscape: on the one side the desert, of smooth and spotless palest fawn-colored sand, slightly undulating, and beyond, the Atlantic; on the other, the west, side, a few valleys and hills, *densely* clothed with a short, almost moss-like (to look down at) growth of huckleberry, blue-berry, bear-berry, josh-pear (which is so abundant in Provincetown), bayberry, rose, checker-berry, and other bushes, and beyond, the Bay. All these bushes formed an even and dense covering to the sand-hills, much as bear-berry alone might. It was a very strange scenery. You would think you might be in Labrador, or some other place you have imagined. The shrubbery at the very summit was swarming with mosquitoes, which troubled me when I sat down, but they did not rise above the level of the bushes. (*J* IX:452)

Characteristically, this homecoming opens to the unknown. The remarkable landscape of contrasts echoes the passage we have read closely in "The Beach" (*CC* 48). Moreover, the familiar, dense, short plants form an abundant, variegated carpet, as they do in "The Sea and the Desert" (153). Here, however, the carpet suddenly resembles the tundra of Labrador, a "very strange scenery." Or the vision could encompass "some other place," unnamed and imagined. Like the description of kelp from some unnamed ship, the possibilities multiply within the verbal transformations of perspective and scenery. And yet, the walker notes, we still have shrubbery at the summit "swarming with mosquitoes," that most familiar of companions.

If the 1857 excursion is a homecoming and a return, it is also a new exploration. It is certainly the longest walk on Cape Cod that Thoreau undertakes—the kind of excursion we would call a backpacking trip or hike. For most of the journey, he uses map and compass, avoiding the main roads and paths and finding his own way. The walker encounters the Cape with a mixture of recognition and reflection, but he is continually opening his perceptions to new perspectives, new knowledge, and new imaginings. When he revisits familiar creatures like the poverty grass and the upland plover, he discovers new details and offers new ideas. These discoveries resonate strongly with the kinds of botanical preparations Thoreau makes in the summer of 1851—the ways, for example, in which Norway cinquefoil can spark his imagination into new forms of plant-thinking. In a similar vein, when he revisits the shrubby plain near Provincetown, it becomes a magical place, transformed into a northern tundra, the strange scenery of Labrador, or even some further imagined place. These layers of experience in the 1857 excursion show the successive, revisionary patterns in Thoreau's imaginative thinking, and they suggest that plant-thinking is fundamental to the practices Thoreau uses to spark his writing. The figures in Thoreau's botanical carpet never stop growing, even if *Cape Cod* is trimmed at the edges by the writer's premature death.

CHAPTER 3

Walden as Botanical Excursion

One of the most fascinating stories about the writing of *Walden* relates to the several revisions Thoreau made to his masterpiece, especially in the years 1852–54. The story has been told by several eminent scholars, among them J. Lyndon Shanley, Ronald A. Clapper, Robert Sattelmeyer, Stephen Adams and Donald A. Ross, and, most recently, Laura Dassow Walls.[1] Given the botanical focus of my readings, the early drafts and the later additions and revisions can be detected, at least in part, by the way Thoreau treats plants in the text. In the later revisions and later chapters of the book, Thoreau develops the plant-thinking version of dynamic, vitalistic writing. The plants are an observational record, but they also provide a way for the writer to find and develop a regenerative vision. This is a new way of seeing the botanical excursion, as a new way for Thoreau to see his work. It is not so much that *Walden* should be read as a botanical excursion, although doing so adds a dimension to our reading of it; the stakes are not high in terms of genre or even subgenre. In a more fundamental way, however, plants are a key to Thoreau's seasonal vision, and they lead him to phenology as one way of interpreting nature, time, and human culture. As we see in Walls's biography, a signal moment for Thoreau occurs in November 1851, when, after a two-year hiatus, he picks up the manuscript of his book and breathes new life into the project. In the final phase of writing his masterpiece, Thoreau finds a new use for the journal entries, plants, and plant-thinking. The new use is an act of renewal and revisioning, transforming Thoreau's draft into the final version of *Walden*.

CLOSE TO HOME

Unlike the botanical excursions of *The Maine Woods* and *Cape Cod,* the travel in *Walden* remains strictly local. "I have travelled a good deal in Concord," pronounces the narrator in the third paragraph of "Economy," and "the inhabitants have appeared to me to be doing penance in a thousand remarkable ways" (*W* 4).[2] The travel is mainly by foot, with occasional forays in boat or on skates. Indeed, the narrator makes yet another pronunciamento about such travel: "I have learned that the swiftest traveller is he that goes afoot" (53). He claims that he can travel to Fitchburg, or even more exotic places, more swiftly and economically than a person who earns the wages for a railroad ticket. The narrator of "Economy" holds consistent opinions. In *A Week on the Concord and Merrimac Rivers,* he had claimed that "the cheapest way to travel, and the way to travel the furthest in the shortest distance, is to go afoot, carrying a dipper, a spoon, and a fish-line, some Indian meal, some salt, and some sugar" (*Wk* 305). In both passages, Thoreau seems to be insisting on travel in figurative senses. The swiftness is more than literal speed, encompassing a sense of ease. The word "furthest," as opposed to the physical distance implied by "farthest," suggests a metaphorical journey, one that can go "furthest" even in "the shortest distance." That last suggestion accords well with the idea of local excursions, on foot.

Literal speed and distance, of course, are not really Thoreau's point at all. His travel is instead deep, an attempt to go "furthest." As the Journal of 4 December 1856 intimates, plants are often the reason for the local travel: "I often visited a particular plant four or five miles distant, half a dozen times within a fortnight, that I might know exactly when it opened" (*J* IX:158). In *Walden,* the visits are not so rapid and persistent. In the late chapter "Former Inhabitants; and Winter Visitors," for instance, Thoreau maps his excursions through "the deepest snows" as a "meandering dotted line," but "no weather interfered fatally with my walks, or rather my going abroad, for I frequently tramped eight or ten miles through the deepest snow to keep an appointment with a beech-tree, or a yellow-birch, or an old acquaintance among the pines" (*W* 265). To "go abroad"—ironically echoing the phrase for a journey to a distant land—is to travel in unmarked paths, to break a trail in deep snow, to "keep an appointment" with a distant tree.

The winter excursions to keep an appointment resonate deeply with an earlier, more developed passage concerning the value of local travel and local plants. In the opening paragraph of "Baker Farm," Thoreau evokes his rambling botanical excursions, even as he details the divinities lurking in the woods and swamps around Concord:

> Sometimes I rambled to pine groves, standing like temples, or like fleets at sea, full-rigged, with wavy boughs, and rippling with light, so soft and green and shady that the Druids would have forsaken their oaks to worship in them; or to the cedar wood beyond Flint's Pond, where the trees, covered with hoary blue berries, spiring higher and higher, are fit to stand before Valhalla, and the creeping juniper covers the ground with wreaths full of fruit; or to swamps where the usnea lichen hangs in festoons from the black-spruce trees, and toad-stools, round tables of the swamp gods, cover the ground, and more beautiful fungi adorn the stumps, like butterflies or shells, vegetable winkles; where the swamp-pink and dogwood grow, the red alder-berry glows like eyes of imps, the waxwork grooves and crushes the hardest woods in its folds, and the wild-holly berries make the beholder forget his home with their beauty, and he is dazzled and tempted by nameless other wild forbidden fruits, too fair for mortal taste. Instead of calling on some scholar, I paid many a visit to particular trees, of kinds which are rare in this neighborhood, standing far away in the middle of some pasture, or in the depths of a wood or swamp, or on a hill-top; such as the black-birch, of which we have some handsome specimens two feet in diameter; its cousin the yellow-birch, with its loose golden vest, perfumed like the first; the beech, which has so neat a bole and beautifully lichen-painted, perfect in all its details, of which, excepting scattered specimens, I know but one small grove of sizable trees left in the township, supposed by some to have been planted by the pigeons that were once baited with beech nuts near by; it is worth the while to see the silver grain sparkle when you split this wood; the bass; the hornbeam; the *Celtis occidentalis*, or false elm, of which we have but one well-grown; some taller mast of a pine, a shingle tree, or a more perfect hemlock than usual, standing like a pagoda in the midst of the woods; and many others I could mention. These were the shrines I visited both summer and winter. (W 201–2)

This paragraph is composed of only three sentences, although the first two sentences are lengthy catalogs of places and trees, respectively. In the first sentence, Thoreau lists ten species of vascular plants and the usnea lichen, as well as toadstools and other beautiful fungi. He is especially taken with the flowering and fruiting plants of the swamp, emphasizing the magical quality of the place; more important than the plants themselves, it seems, is the effect they have on the "beholder," who, bewitched, "forgets his home with their beauty." Indeed, for modern readers some of the names that Thoreau uses are misleading, if not inaccurate. The "dogwood," for instance, is *Rhus venenata* in Gray's *Manual*, also known as poison sumac. The "red alder-berry" belongs to the species *Ilex verticillata*, first identified by Gray and described as having "bright red" fruit; its common name, "black alder," is less common now than "winterberry." The "creeping juniper" may be a prostrate form of common juniper, but the name Thoreau uses is perfect in the context of the description. After all, Thoreau is intent on creating a mythic, somewhat creepy landscape, with the allusions to the Druids, the Norse hall of heroes slain in battle, and the "imp" of folklore. This homespun, pagan language accords well, moreover, with naming Flint's Pond and using the most common names for the plants and trees. Who needs to travel abroad when there are "fleets at sea" on the way to Flint's Pond?

In the second catalog, the rarity of Thoreau's favorite trees marks them as deserving of our worshipful attention. These trees grow in out-of-the-way places, requiring special effort on the part of the human visitor. As we saw earlier, the beech and yellow birch are local rarities, calling for visits even in the deepest snow. In the "Baker Farm" passage, Thoreau gives the beech a more exotic air by making the last remaining grove the result of beechnuts used to bait "pigeons." He is certainly referring to passenger pigeons, hunted to near extinction in the Concord area by the 1850s. Thoreau had seen and heard these wild pigeons his entire life, but, as he notes in the Journal of 9 May 1852, "Saw pigeons in the woods with their inquisitive necks and long tails—but few representatives of the great flocks that once broke down our forests" (*PJ* 5:49). One wonders if the "small grove of sizable trees" was planted by carcasses or by living birds that escaped the slaughter. At first the list of trees requires a loving description for each species, but by the end of the sentence Thoreau is simply listing them, as if the common names can evoke the plants' uncommon beauty. Then he inserts the only scientific name in the catalog: *Celtis occidentalis*. This is the common hackberry, a tree familiar to Thoreau from his youth, but the scientific name may evoke the fact that Concord now has "only one well-grown," as if an uncommon specimen requires an uncommon name. The rare trees are towering specimens, and for that reason the narrator compares them to temples, pagodas, and shrines.

Thoreau's extended catalogs emphasize specific plants in the neighbor-hood, but the paragraph does not represent a new way of thinking about the relationship of the writer to the plants. Instead, the passage functions most fundamentally as an extended metaphor. The places and plants be-come representative of the magical, otherworldly, and spiritual qualities of the local woods and swamps. The descriptions of specific plants support the extended figuration of worship. Thoreau uses visual images to emphasize the beauty of the trees: the "loose golden vest" of the yellow birch; the "beautifully lichen-painted" trunk of the beech; the "silver grain" sparkling when you split the wood.

The visual aesthetics of trees and other plants dominates the first pas-sage cataloging plants in *Walden*. It is a playful remark whose source comes from an early Journal entry, written during the 1845–47 period at Walden Pond: "I have watered the red-huckleberry & the sand cherry— and the hopwood-tree—& the cornel—& spoonhunt—and yellow violet which might have withered else—in dry seasons. The white grape" (*PJ* 2:228). The revised passage appears in "Economy," in a sequence of humorous descrip-tions of the narrator's many occupations. The paragraph begins with the narrator's work of looking after "the wild stock of the town" and then leads to his repeated virtuous actions: "I have watered the red huckleberry, the sand cherry and the nettle tree, the red pine and the black ash, the white grape and the yellow violet, which might have withered else in dry seasons" (*W* 18). The names of some of the plants in the catalog are no longer com-mon, although most of the plants can be found in Gray's *Manual of Botany* and in other passages from the Journal.[3]

In the Journal entry of 1845–47, "hopwood tree" and "spoonhunt" are both unfamiliar names, though they do provide assonance and alliteration to the list. They are deleted from the version in *Walden*. In the published ver-sion, colors dominate the catalog, giving the plants a visual beauty even as the narrator splashes them. Moreover, the list composes itself into doublets, a kind of balanced, nearly metrical procession of colorful names. "Nettle tree" is no longer a common name; interestingly, it is another term for *Celtis occidentalis,* the common hackberry listed in the "Baker Farm" passage. It is paired with the sand cherry, an abundant plant that Thoreau names in other places, both in the Journal and in *Walden*. In the "Sounds" chapter, for instance, he includes it in a long list of plants growing in the front yard of his house. He names the "sand-cherry" as "*Cerasus pumila,*" a term he takes from André Michaux (*W* 113).[4]

THE CONNECTING LINK

In a host of significant ways, "The Bean-Field" suggests the importance of Thoreau's botany to a reading of *Walden*. The chapter is part of the narrative from early on, just as it narrates most tellingly Thoreau's labor as a farmer during the first year at Walden Pond. Even before he took up residence at the pond in 1845, Thoreau had planted some of the beans. In a Journal entry from 7 July 1845, he reflects on the place he occupies and links his role as a heroic traveler to the beans and to a pitch pine:

> I am glad to remember tonight as I sit by my door that I too am at least a remote descendent of that heroic race of men of whom there is tradition. I too sit here on the shore of my Ithaca a fellow wanderer and survivor of Ulysses. How Symbolical, significant of I know not what, the pitch pine stands here before my door unlike any glyph I have seen sculptured or painted yet— One of nature's later designs. Yet perfect as her Grecian art. There it is, a done tree. Who can mend it? (*PJ* 2:156)

The pitch pine, in resisting Thoreau's efforts to make it "Symbolical," opens the writer to a rich encounter. In reflecting on himself as a "survivor of Ulysses," the writer casts his eyes about him, searching for a correlative image. The tree is as significant and "perfect" as Grecian art, but how it signifies or what it signifies is left open. Fortunately, Thoreau does not link the pine tree to Ulysses's mast, and, instead of lapsing into insignificance or a lack of meaning, the pitch pine stands before the writer's door, ready for further imaginative encounters. Unable to create an immediate correspondence between the tree and his imagination or to "mend" the perfection of the "done tree," Thoreau's mind moves to the ideas of the present generation of heroes, the remains of Native Americans, and then to "the Great spirit." A vital, dynamic spiritual reality is always close by: "But nearest to all things is that which fashions its being. Next to us the grandest laws are being enacted and administered" (*PJ* 2:157). The pitch pine leads Thoreau to heroic thoughts of "Grecian art" and what we leave behind, to the "Great spirit" whose creations we creatures are.[5] Then he turns his thoughts to the bean field:

> My auxiliaries are the dews and rains—to water this dry soil—and genial fatness in the soil itself, which for the most part is lean and effoete. My enemies are worms cool days—and most of all woodchucks. They have nibbled for me an eighth of an acre clean. I plant in faith—and they reap—this is

the tax I pay—for ousting Jonswort & the rest. But soon the surviving beans will be too tough for woodchucks, and then—they will go forward to meet new foes. (2:159)

Thoreau ends the opening paragraph of "The Bean-Field" with a revised version of this passage from the Journal. Both versions use the military term "auxiliaries," meaning "foreign or allied troops used in the service of a nation at war" (OED). Along with the word "enemies," the term connects the beans to the heroic opening of the Journal entry. The revisions at first seem less epic. The "genial fatness" of the soil becomes "fertility." The eighth of an acre becomes a quarter of an acre. Two sentences ("I plant in faith, and they reap. This is the tax I pay for ousting Jonswort and the rest.") become one question: "But what right had I to oust johnswort and the rest, and break up their ancient herb garden?" (W 155).[6] The question relates the narrator to the woodchucks in terms of rights, replacing the competitive, inimical language of the Journal sentences, with their imagery of the woodchucks reaping what the writer sows and taxing him for weeding out the wild plants. The bean battlefield becomes "their ancient herb garden."[7]

This initial revision is immensely productive. The chapter turns on the questions of value, of right, and of labor. In a particularly salient passage, Thoreau addresses these questions as directly as he can:

This was one field not in Mr. Colman's report. And, by the way, who estimates the value of the crop which Nature yields in the still wilder fields unimproved by man? The crop of *English* hay is carefully weighed, the moisture calculated, the silicates and the potash; but in all dells and pond holes in the woods and pastures and swamps grows a rich and various crop only unreaped by man. Mine was, as it were, the connecting link between wild and cultivated fields; as some states are civilized, and others half-civilized, and others savage or barbarous, so my field was, though not in a bad sense, a half-cultivated field. They were beans cheerfully returning to their wild and primitive state that I cultivated, and my hoe played the *Ranz des Vaches* for them. (W 158)

This may be Thoreau's most cheerful aside. The combination of the commissioner of agriculture and the song of Swiss cowherds is already humorous, and even Thoreau's version of the hierarchy of civilizations seems to make fun of the clichés of progress. The field resists the structured closing down of meanings, moving in the direction of "the still wilder fields unimproved by man." Thoreau insists that his field is only "half-cultivated," and that makes it "the connecting link between wild and cultivated fields." In the context of "The Bean-Field," Thoreau cultivates the soil by tilling it; he cultivates the plants with his hoe, improving them for commercial purposes. Of course, he is playing upon the other figurative meanings of "cultivate," as improving a mind or a person or a village or a country through education or training. These meanings were at work in the English language from the first uses of the word in the seventeenth century (OED); they coexist in the usages of the word "cultivate," even today, and they keep the questions of value, right, and labor in play every time anyone uses the word.

Thoreau develops the meanings of "cultivate" further in another salient passage from "The Bean-Field," in a paragraph that develops the art of cultivating an intimate acquaintance with another being. In Thoreau's case, the other being is a plant, or a number of plants: "It was a singular experience that long acquaintance which I cultivated with beans, what with planting, and hoeing, and harvesting, and threshing, and picking over, and selling them,—the last was the hardest of all,—I might add eating, for I did taste" (W 161). The "singular" becomes plural, what with all the necessary parts of cultivating the beans, let alone cultivating their acquaintance. And yet it seems that the acquaintance is most valuable, for Thoreau even cultivates an acquaintance, "intimate and curious," with weeds: "That's Roman wormwood,—that's pigweed,—that's sorrel,—that's piper-grass,—have at him, chop him up, turn his roots upward to the sun, don't let him have a fibre in the shade, if you do he'll turn himself t'other side up and be as green as a leek in two days" (161). The list echoes a previous passage in the chapter, in which Thoreau remarks that the beans are themselves a kind of weed, and that he is choosing to make "the yellow soil express its summer thought in bean leaves and blossoms rather than in wormwood and piper and millet grass, making the earth say beans instead of grass" (157). This is a memorable moment in Walden: the plants are words spoken by the soil, expressions of the earth and its summer thought. The plants link the physical labor of cultivation with a cultivating, productive language, perhaps most closely linked to what Thoreau calls, in "Sounds," the "language which all things and events speak without metaphor, which alone is copious and standard" (111). As a useful gloss on the language "without metaphor," consider this passage from the Journal of 23 August 1845: "In all the dissertations—on language—men forget the language that is—that is really universal—the inexpressible meaning that is in all things & every where with which the morning & evening teem. As if language were especially of the tongue. Of course with a more copious hearing or understanding—of what is published the present *languages* will be forgotten" (*PJ* 2:178).

"The Bean-Field" not only represents a connecting link between wild and cultivated fields; it creates that connecting link, repeatedly and consciously. The writer cultivates weeds at least as assiduously as he does his beans, in the sense that he cultivates their acquaintance and deepens his knowledge of them in the process. The beans are simple, a single species of "the common small white bush bean" (*W* 163). The wild weeds, on the other hand, are multitudinous and various, each one claiming its identity in the list of four species: "That's Roman wormwood,—that's pigweed,—that's sorrel,—that's piper-grass" (161). These are not valuable plants for commercial purposes, but Thoreau values them for their names and their identities. They are part of the "rich and various crop only unreaped by man," growing in wild places around Concord and deserving, in their own way, as much care as the "crop of *English* hay" so carefully weighed, measured, and calculated (158). The value of the crop may be less than the value of cultivating the crop, no matter what the financial benefits or costs may be. So, for instance, Thoreau can famously claim that "some must work in fields if only for the sake of tropes and expression, to serve a parable-maker one day" (162).

Though it is right to search the pages of *Walden* for parables, it is also worth pausing to see how valuable unreaped crops can be. Thoreau raises beans only during the first year at Walden Pond, claiming that he then attempts to plant the seeds of virtues like "sincerity, truth, simplicity, faith, innocence, and the like," to see if they will grow and sustain him. Over several years, right up to the publication of *Walden,* the seeds have not borne fruit, "were wormeaten or had lost their vitality, and so did not come up" (164). If the metaphor serves, it shows how difficult it is to cultivate any virtue. Thoreau takes on the failure as his own, but in the rest of the chapter it becomes more general. Time and again, our acquaintances focus on beans and corn, not on invisible seeds or unreaped crops. Even the beans that Thoreau cultivates bear other crops he will never harvest: "These beans have results which are not harvested by me. Do they not grow for woodchucks partly?" (166). Here Thoreau returns to the Journal entry that serves as the seed for the whole chapter. The harvest is greater than an ear of wheat, the "hope of the husbandman" (166), and for that reason the harvest can never fail.

More than the connecting link between wild and cultivated fields, "The Bean-Field" connects material things and words, natural objects and virtues, botanical growth and a faith in personal and artistic growth. In Thoreau's plant-thinking, moreover, faith grows, like a seed, and like other virtues. The harvest also raises the theme of use: the thoughts on the rights of the woodchuck to part of the bean crop recall the meditation on the highest use of the pine tree in "Chesuncook." In both works, Thoreau is considering the way human beings make use of plants and trees. His views in "Chesuncook" favor the poet over the lumberman, and in "The Bean-Field" he gives a part to the woodchuck. Indeed, the word "use" may itself be misapplied in the case of the bean field. With a more copious understanding, we might instead think of "results" beyond our immediate needs and perceptions: the results could be virtues, or they could be the full bellies of woodchucks. In either case, Thoreau suggests, thinking *with* the plants can open our perceptions to new and surprising possibilities of meaning.

HALF-CULTIVATED FIELDS

A connecting link implies that both sides of the link are necessary. The "half-cultivated field" is also half-wild; it is a meeting ground for a host of apparently irreconcilable oppositions that we could range under the headings "wild" and "cultivated." In the meetings, moreover, the field becomes a ground for new possibilities, new perceptions. For these reasons, it is valuable to bear in mind that Thoreau remains in practical contact with the earth. When he cultivates his beans, he means what he says and points in all the directions his words indicate. The lists of expenses and profits in the "Economy" and "Bean-Field" chapters are as meaningful, in their way, as the lists of plants in "Baker Farm" or the seeds of virtues that Thoreau lists in "The Bean-Field." When we try to read *Walden* as a transcendentalist or materialist or environmentalist document, we are always missing some direction we might take. In this sense, the book is an excursion, for it requires that readers remain in motion, never settling too easily into one field of meaning. Because plants have opened this point, too, we can see *Walden,* at least provisionally, as a botanical excursion.

Thoreau's botanical excursions often yield practical results, but rarely do they stop there. He opens the chapter "House-Warming," for example, with a long paragraph that is reminiscent of the "Baker Farm" hymn to the shrines in the woods. October is the month for harvests, but many of the writer's fields remain unreaped:

In October I went a-graping to the river meadows, and loaded myself with clusters more precious for their beauty and fragrance than for food. There too I admired, though I did not gather, the cranberries, small waxen gems, pendants of the meadow grass, pearly and red, which the farmer plucks with an ugly rake, leaving the smooth meadow in a snarl, heedlessly measuring them by the bushel and the dollar only, and sells the spoils of the meads to Boston and New York; destined to be *jammed,* to satisfy the tastes of lovers of Nature there. (*W* 238)

As unbendingly virtuous as Thoreau can be, the passage does not present the harvest as a choice between poetry and lucre. In fact, the writer is loading himself with clusters of grapes, and his haul is no less prodigious for its beauty and fragrance. He does not gather the cranberries in October, but that may simply be because he is waiting for a frost to deepen the sweetness of the fruits. His fundamental critique is against the farmer's methods and the scale of his harvest. The cranberries are too beautiful to become *jammed,* but that is their destiny.

As the paragraph continues, Thoreau lists other, further harvests. He lets the "barberry's brilliant fruit" stay on the plants, but he gathers "a small store of wild apples for coddling" and "half a bushel" of ripe chestnuts for winter. The chestnuts are so tempting that he cannot wait for the frost, and he admits to stealing "half consumed nuts" from the red squirrels and blue jays. Like the woodchucks, the squirrels and jays get most of the harvest near Thoreau's house, but he ventures farther afield to find a larger crop. He promotes chestnuts as a substitute for bread, and he adds another possibility with the "ground-nut (*Apios tuberosa*), on its string, the potato of the aborigines, a sort of fabulous fruit" (*W* 239). In an earlier allusion to "butchers rak[ing] the tongues of bison out of the prairie grass" (238) and in the apparent aside that "cultivation has well nigh exterminated" the ground-nut, Thoreau suggests that the genocide of Native Americans reaches across the continent. The chestnut woods were also formerly "boundless," but now the trunks "sleep their long sleep under the railroad" (238).

The botanical excursions create many harvests, not least in the writer's vision of a possible future. Thoreau imagines "wild Nature" being allowed to "reign here once more," and all the cultivated, "tender and luxurious English grains will probably disappear before a myriad of foes." Even "the last seed of corn" might return to the American Southwest, leaving New England to the humble ground-nut, to "revive and flourish in spite of frosts and wildness, prove itself indigenous, and resume its ancient importance and dignity as the diet of the hunter tribe" (W 239). In the spirit of "Chesuncook," moreover, the future importance of the ground-nut and the revival of the Indigenous tribes of New England are linked to a new "reign of poetry" in the land.

Among the possibilities that derive from plant-thinking, picking huckleberries is perhaps the most likely. As we have noted, Thoreau entertains the idea of picking huckleberries as a way of making his living. He would make his occupation of "ranging the hills all summer to pick the berries which came in my way, and thereafter carelessly dispose of them" (W 69–70). He even dreams big dreams of gathering "the wild herbs," or carrying "evergreens to such villagers as loved to be reminded of the woods, even to the city, by hay-cart loads." But "trade curses every thing it handles," and the business of berry-picking is not to be.

Thoreau revisits the idea of berry-picking in the opening of "The Ponds," and the work is once again inimical to commercial purposes. He rambles far afield, making his supper of "huckleberries and blueberries on Fair Haven Hill," but no purchaser of the fruits, and no farmer raising them for market, can possibly taste their "true flavor." Only by picking the berries yourself can you taste them. The point takes on comic proportions: "A huckleberry never reaches Boston; they have not been known there since they grew on her three hills. The ambrosial and essential part of the fruit is lost with the bloom which is rubbed off in the market cart, and they become mere provender. As long as Eternal Justice reigns, not one innocent huckleberry can be transported thither from the country's hills" (W 173). Like poetry, Eternal Justice will always reign!

In one harvest, at least, Thoreau seems in good earnest. In "The Ponds," he seriously laments the logging of the "thick and lofty pine and oak woods" around Walden Pond, and in retrospect he sees that the "wood-choppers have still further laid them waste, and now for many a year there will be no more rambling through the aisles of the wood, with occasional vistas through which you see the water" (W 191, 192). At the same time, he celebrates his excursions to gather dead wood for his fireplace. In "House-Warming," he details his winter trips to "collect the dead wood in the forest, bringing it in my hands or on my shoulders, or sometimes trailing a dead pine tree under each arm to my shed" (W 248). He delights in "stealing" an "old forest fence" for fuel and clearing out "fagots and waste wood" that "some think, hinder the growth of the young wood." And there is the driftwood along the pond itself, including a "raft of pitch-pine logs with the bark on" (249). He hauls it up on shore and leaves it to dry for six months. Then he amuses himself "one winter day with sliding this piecemeal across the pond, nearly half a mile, skating behind with one end of a log fifteen feet long on my shoulder, and the other on the ice" (249). After such efforts, it is no wonder the narrator can assert, "Every man looks at his wood-pile with a kind of affection" (251).

The loving account of this harvest runs for several paragraphs, ending only with one of Thoreau's best poems, "Smoke":

Light-winged Smoke, Icarian bird,
Melting thy pinions in thy upward flight,
Lark without song, and messenger of dawn,
Circling above the hamlets as thy nest;
Or else, departing dream, and shadowy form
Of midnight vision, gathering up thy skirts;
By night star-veiling, and by day
Darkening the light and blotting out the sun;
Go thou my incense upward from this hearth,
And ask the gods to pardon this clear flame.[8]

Perhaps because the poem settles into an unrhymed pentameter, it has the quality of a brief, dignified prayer. The metrical substitutions, especially the initial trochees and dactyls in several lines, show considerable skill. The flair for dramatic, fluent phrasing is balanced by a humble search for fitting imagery. No mention of plants, but we hardly miss them.

SAND FOLIAGE AS CONNECTING LINK

Many readers may consider the sand foliage passage in "Spring" as the climax of the second half of *Walden*.[9] For the significance of Thoreau's botany, sand foliage functions in two important ways. First, it demonstrates the successive, overlapping strategies of revision that mark Thoreau's botanical excursions, bringing *Walden* into close connection with *The Maine Woods* and *Cape Cod*. Second, it demonstrates the fundamental role of plant-thinking in Thoreau's work of the 1850s. Adams and Ross support the first claim by asserting that some twenty-three passages from the Journal appear in the four paragraphs of the sand foliage passage.[10] My research suggests that Thoreau first began writing about the sand foliage in the Deep Cut in the spring of 1848, after he had left Walden Pond. In the next six years, he returned to the phenomenon repeatedly, adding to the descriptions and meanings of the thawing clay and sand bank. Ultimately, however, the sand foliage is more than a passage of accretion; it is a passage to transformation.

The two versions from the spring 1848 period are quite close to one another, and they give the initial outline of the entire four-paragraph description. The two are in fact the first and second draft of the sand foliage passage, revised and reordered. The first entry is comprised of twelve "paragraphs" or segments, marked by indentation (*PJ* 2:382–84). The later (undated) revision numbers the segments in the margin, from 1 to 17, indicating an order of paragraphs. With some related numberings given single or double slashes (9 and 9', 14, 14', and 14''), the later revision is comprised of twenty segments, not all marked by indentation (2:576–78). An illustration showing a sample page of the manuscript (facing 2:455) gives a good idea of how the two drafts form a palimpsest or hybrid text.

The revisions show Thoreau's rigorous attention to the order of ideas, an order that persists in the final version in *Walden*. The first of the four paragraphs in the published version is devoted to describing the forms of thawing sand and clay as they flow down the sides of the Deep Cut. The second begins by naming the "foliage, or sandy rupture," and then it moves to the effects the flowing sand foliage has upon the observer, including the ideas that arise in his mind. By the end of the second paragraph, Thoreau can assert that "the whole tree is but one leaf, and rivers are still vaster leaves" (307), a line that has its source in the 1848 Journal. In the third paragraph, Thoreau develops further ideas about the sand foliage as a pattern for physical nature in all its dynamic manifestations. Thus the foliage models "how blood vessels are formed." Ultimately a human being can be seen as "a mass of thawing clay," developing from "drop-like point" to "channel or artery" to "fleshy fibre or cellular tissue" (307–8). Again, the first formulation of these ideas occurs already in the 1848 Journal (*PJ* 2:383, 577–79).

The fourth paragraph of the sand foliage passage leads with a sentence that forms the conclusion to the revised 1848 draft: "Thus it seemed that this one hillside illustrated the principle of all the operations of Nature" (*W* 308). The published sentence echoes the first draft and revision, though it revises them as well: "So it seemed as if this one hill side contained an epitome of all the operations in nature" (*PJ* 2:384). The revision keeps the wording of the first draft exactly, but it positions the sentence as the conclusion of the sand foliage passage (2:578). In the published version, Thoreau changes "contained an epitome" to "illustrated the principle," making the sentence more directly active, powerful, and universal. Instead of a conclusion, moreover, the sentence leads to a highly developed final paragraph of implications and insights, the most famous of which is probably "There is nothing inorganic" (W 308).

From the 1848 beginnings, the four-paragraph sand foliage passage in *Walden* grows, and each of the four paragraphs develops and deepens the initial drafts in important ways. The Journal of December 1851 to April 1852 does indeed give many sand foliage entries, some of them quite brief. The entries of 31 December 1851, 10 March 1852, and 15 March 1852 are especially significant for the material and thought they provide. Four entries from 1854 are equally important: 5 February, 8 February, 2 March, and 5 March. All seven of these reflections find expression in the published version. In addition, however, the published version includes significant material for which I have found no clear source in the Journal.

On 31 December 1851, Thoreau begins by lamenting that he is too late to see the sand foliage. Still, he finds "in some places it is perfect. I see some perfect leopard's paws," an image that finds a place in the first paragraph of the published text. More important than the image, however, is the idea of eternal motion and life within the earth: "I seem to see some of the life that is in the spring bud & blossom more intimately nearer its fountain head— the fancy sketches & designs of the artist" (PJ 2:230). In a surprising idea, the "forms of foliage" may come "before plants were produced to clothe the earth." Therefore, Thoreau concludes, the earth is "not a dead inert mass. It is a body—has a spirit—is organic—and fluid to the influence of its spirit—and to whatever particle of that spirit is in me" (2:230). The primary connecting link here is body and spirit, forming an organic whole, but Thoreau also connects his own "particle of that spirit" to the earth, imagining that the earth is "fluid to the influence" of both spirits. The "artist" is spirit, in whatever form it takes. While the 1848 Journal had struck upon the image of the "studio of the artist," here the artist takes on a more defined power.

For Thoreau, this inner spirituality is by no means ethereal. As a source of growth and life, it displays "fertility & luxuriance," and it is also "somewhat foecal and stercoral." The site of fecundity is very close to the site of excretion: "There is no end to the fine bowels here exhibited—heaps of liver-lights & bowels. Have you no bowels? Nature has some bowels. [A]nd there again she is mother of humanity" (PJ 2:231). "Liver-lights" refers to the combination of liver and lungs, often used for cat and dog food nowadays. Thoreau is being both jocular and challenging when he asks, "Have you no bowels?"[11] This passage appears in the fourth paragraph of the published version, summarizing the effect the sand foliage has on the narrator: "This phenomenon is more exhilarating to me than the luxuriance and fertility of vineyards. True, it is somewhat excrementitious in its character, and there is no end to the heaps of liver lights and bowels, as if the globe were turned wrong side outward; but this suggests at least that Nature has some bowels, and there again is mother of humanity" (W 308).

Plant-thinking is suggested in the first drafts in 1848, and it develops in a number of directions through the entries of 1851–52 and 1854. For example, Thoreau's original idea of the "mythological vegetation" (PJ 2:383) becomes the more elaborate revision: "Little streams & ripples like lava over flow & interlace one with another like vines—a sort of Hybrid product—obeying half way the law of currents & half way the law of vegetation. As it were mythological vegetation or like the forms which I have seen imitated in bronze" (2:577). In Walden, Thoreau keeps the idea of the "hybrid product" that "obeys half way the law of currents, and half way that of vegetation," but the "mythological vegetation" becomes "a truly grotesque vegetation, whose forms and color we see imitated in bronze, a sort of architectural foliage more ancient and typical than acanthus, chiccory, ivy, vine, or any vegetable leaves" (305). "Architectural foliage" is a first phrase for the "hybrid product" of currents and vegetation; "sand foliage" will be the second, in the following paragraph (306). The "architectural foliage" and "grotesque vegetation" both allude to Greek and Roman decorative motifs, especially the scrolling foliage on capitals, columns, friezes, and sculptures. Thoreau refers to the "architectural foliage" of the Deep Cut as "more ancient and typical than acanthus, chiccory, ivy, vine, or any vegetable leaves" (305); this list of botanical motifs derives from the history of Greek decorative arts. In the Journal of 26 January 1852, he makes an initial entry for the botanical list: "In these fresh designs there is more than the freedom of Grecian art— more than acanthus leaves" (PJ 4:294). The foliage created at the Deep Cut is more ancient than Greek art and architecture, and it stands as a primary model for the "grotesque" in art history, the painting and sculpture that consists of "representations of portions of human and animal forms, fantastically combined and interwoven with foliage and flowers" (OED). Thoreau even imagines that the forms of sand foliage might become "a puzzle to future geologists" more than to art historians or literary scholars (305). At its source, the "hybrid product" of sand foliage includes currents and vegetation, but it extends to include art and nature, literature and architecture, classical decorations and an ever-shifting, present design, freer and fresher than any botanical motif in stone or bronze. The design, springing out of the railroad cut, presents leaves "before any vegetable leaves."

If the Deep Cut takes plant-thinking in the direction of a fresh and free architectural foliage, it does not neglect to create forms of real plants and allies. On 15 March 1852, Thoreau observes the sand flowing in the Cut and remarks, "These heaps of sand foliage remind me of the laciniated—lobed—& imbricated thallusses of some lichens—somewhat linear laciniate" (*PJ* 4:388). The entry finds a nearly perfect echo in the sand foliage passage: "As it flows it takes the forms of sappy leaves or vines, making heaps of pulpy sprays a foot or more in depth, and resembling, as you look down on them, the laciniated lobed and imbricated thalluses of some lichens" (*W* 305). Thoreau is employing botanical vocabulary with great precision here. The thallus of a lichen is the main body; the lobes on a lichen indicate a common leaf-like growth form called *foliose*. "Laciniate" is a botanical term describing a margin of a leaf, petal, or lobe as divided into segments, often long and narrow ("linear") or overlapping ("imbricate"). A synonym for "laciniated" would be "fringed."[12] The resemblance of the sand foliage to lichens is especially telling because both are "hybrid products" and present a complex interweaving of life forms or laws. Lichens are composite organisms, in which algae and cyanobacteria form a symbiotic relationship with a fungus, living within the structure provided by the filaments (*hyphae*) of the fungus.[13]

The four paragraphs of the sand foliage passage are themselves "hybrid products," made up of imbricated Journal entries and drafts from other sources. As a text, the sand foliage is less a product than a process, and for that reason the paragraphs shift and move as we read them. We can find sources for many of the individual words and sentences in the published version, but the sources do not show how Thoreau's imagination puts the pieces together. In the first paragraph, for example, we have seen many sources from the Journal, but the final version in *Walden* includes several ideas and images that are not readily found in the Journal. We can find parts of the paragraph stretching all the way back to the 1848 Journal and all the way forward to March 1854, but these parts do not explain the final, long sentence, in which Thoreau describes the current of the sand foliage forming *strands,* expanses of *sand,* and *banks* before the "forms of vegetation are lost in the ripple marks on the bottom" (*W* 305). Rehearsing the large-scale forms as they flow down the railroad cut, the italicized words somehow echo the *"grotesque* vegetation" while remaining distinct from it.

Thoreau's focus on process relates particularly well to the relationship between the Journal and *Walden*. The sources and their revisions help us understand the ways in which Thoreau views the sand foliage over a period of several years, repeatedly returning to the scene to see it more and more deeply. By reading the different versions in concert, we discover deeper resonances in the published version. In addition, some of the Journal entries stand out as meaningful encounters with the sand foliage, further enriching the published text. A good example of this kind of hybrid, imbricated reading comes in the 5 February 1854 Journal:

> That sand foliage! It convinces me that nature is still in her youth—That florid fact about which Mythology merely mutters—that the very soil can fabulate as well as you or I. It stretches forth its baby fingers on every side. Fresh curls spring forth from its bald brow— There is nothing inorganic— This earth is not then a mere fragment of dead history—strata upon strata like the leaves of a book—an object for a museum & an Antiquarian but living poetry like the leaves of a tree—not a fossil earth—but a living specimen. You may melt your metals & cast them into the most beautiful moulds you can—they will never excite me like the forms which this molten earth flows out into— The very earth—as well as the institutions upon it—is plastic like potters clay in the hands of the artist. These florid heaps lie along the bank like the slag of a furnace—showing that nature is in full-blast within. but there is No admittance except on business. Ye dead & alive preachers Ye have no business here. Ye will enter it only as your tomb. (*PJ* 7:268)

The sand foliage becomes a holistic figure for nature "still in her youth." In Marder's analysis, the plant figures *phusis,* the fundamental and elusive vitality in all of nature and a suitably dynamic synonym for the ontological concept of Being (*Plant-Thinking* 28–29). Even more important for the Journal writer, the foliage is itself capable of fully articulate expression: the "florid fact" precedes the mutterings of mythology, and the very soil can "fabulate as well as you or I." The sand foliage, representing the earth, becomes "living poetry, like the leaves of a tree," a short-circuiting figure of intersecting nature and culture. Moreover, the language of the foliage is molten and plastic, flowing from geology and botany and mythology to metallurgy, sculpture, and pottery within a few sentences. Thoreau ends with the deadly seriousness of this language, created in a fiery furnace that excludes the "dead & alive preachers." He echoes Dante's warning on the gates of Hell, "Ye will enter it only as your tomb."

The 5 February 1854 Journal entry, a summation of years of observation and reflection, delivers Thoreau's plant-thinking and his vision of *phusis* in a highly developed form. The ideas of what Marder calls "vegetal being" and "vegetal existence" ground the enthusiastic encounter with the sand foliage. The "florid fact" begins and then underlies Thoreau's entire meditation on nature, culture, and artistic expression. His vitalist vision, that "there is nothing inorganic," functions as a response to unspoken scientific and philosophical assumptions about the earth. And because the earth is a synonym for nature as *phusis*, we can see that for Thoreau the earth is a form of vegetal being and existence. The "living poetry" of the sand foliage is "like the leaves of a tree" and a "living specimen." It is as if the sand foliage could be collected in Thoreau's botany box hat to become part of his extensive herbarium. Such is the significant extravagance of the Journal passage.

Thoreau's botanical vitalism responds to dead institutions like museums, churches, and schools—to personify them, scientists, preachers, and teachers. To totalize somewhat extravagantly, Thoreau is warning against institutionalized knowledge, embodied in the discourses of science, religion, and the humanities. But Thoreau is also responding to himself and to his own concerns and limitations as a writer. The critique of language is most fundamentally a critique of his own language. We know from the Journal that Thoreau continually prods himself to acquire detailed scientific knowledge about the plants of Concord, and that he becomes a significantly accomplished botanist. We also know that he steps back from his immersion in botany so that he does not succumb to the abstractions and reductions of taxonomy. Even though Thoreau remains a student of Western science, he clearly perceives its limitations and boundaries.

The second paragraph of the published sand foliage passage reveals some of these tensions and their possibilities. Thoreau focuses on the sudden "springing into existence" of the sand foliage, and it affects him "as if in a peculiar sense I stood in the laboratory of the Artist who made the world and me,—had come to where he was still at work, sporting on this bank, and with excess of energy strewing his fresh designs about" (*W* 306). While this moment harks back to the "studio of the artist" of the 1848 versions (*PJ* 2:384, 578) and to the "fancy sketches & designs of the artist" in the 31 December 1851 Journal (*PJ* 4:230), the significant revisions of "studio" to "laboratory," and "artist" to "Artist," suggest a fresh complexity. In 1848, the artist of the fresh designs is God (2:577), but in *Walden* he is less narrowly defined as the Artist. In the fourth paragraph of the published version, Thoreau incorporates most of the 5 February 1854 language, which in turn incorporates the language of the 31 December 1851 Journal. The burden in both Journal entries and in the fourth paragraph is that "there is nothing inorganic," that the earth is "not a fossil earth, but a living earth," whose "great central life" dwarfs all other forms of life and art (*W* 308–9). At one point, the sand foliage convinces Thoreau "that Earth is still in her swaddling clothes, and stretches forth baby fingers on every side" (308). By capitalizing "Earth," Thoreau equates it with "Nature" and allies both figures with "the Artist." All three personifications exhibit "excess of energy," a fundamental vitality that is exhibited in the sudden rupture of the sand foliage into existence, a vitalism that is the source of all creation, material and spiritual.

The fundamental vitalism of Thoreau's botany emerges in at least three ways in the published sand foliage description. First, he develops an origin story of the sand foliage as "an anticipation of the vegetable leaf. No wonder that the earth expresses itself outwardly in leaves, it so labors with the idea inwardly" (*W* 306). Second, he connects the internal and external aspects of the leaf with the constituent elements of language, arguing that the very "radicals" (roots) of words parallel the physical structures of the leaf and its allied words like *lobe, globe, lap, flap,* and *lapse* (306).[14] Third, he uses his observations to develop a theory of physical bodies as constituted by drops, which in turn create the lobes and leaves of plants, human organs and structures, and currents of streams and rivers. Thoreau develops these three elements of the sand foliage paragraphs to suggest how "it seemed that this one hillside illustrated the principle of all the operations of Nature" (308). Plant-thinking is central to this illustration, just as it is central to the writer's unified vision of *phusis*, "the principle of all the operations of Nature."

It is easy to read Thoreau's assertions as mere analogies or rhetorical exaggerations, to infer that he does not mean what he claims in the sand foliage passage. Harder, it is better to take him seriously in his ideas. If there is nothing inorganic, then it is possible to see a unity underlying the parts of earth, including the human body and spirit; one formulation for this unifying principle is earthly vitalism. When Thoreau looks at the sand foliage as an anticipation of the vegetable leaf, for instance, he sees a fundamental vitalism, an excess of creative energy, underlying both phenomena and connecting them in form and motion. When he investigates the relationship between internal and external elements of language, he is arguing fundamentally that there is a vital connecting link of all linguistic elements to one another. When he observes that the flowing sand begins as a "drop-like point" and then forms channels and arteries, connecting the "pulpy leaves or branches" of the foliage, he interprets the physical process as obeying a vital law of self-organization (307). At the very root of the sand foliage passage and its vitalist presentation, moreover, is Thoreau's search for a language to match the sudden, exhilarating emergence of the "foliaceous heaps" (308).

Directly after the hybrid, imbricated summary of "that sand foliage!" in the Journal of 5 February 1854, Thoreau turns to his own language and style of expression. How, he seems to wonder, can he write in a way that will unfold as vitally as the sand foliage or the leaves of plants? His answer comes in the process of answering the warning "Ye will enter it only as your tomb":

> I fear only lest my expressions may not be extravagant enough—may not wander far enough beyond the narrow limits of our ordinary insight & faith—so as to be adequate to the truth of which I have been convinced. I desire to speak somewhere without bounds in order that I may attain to an expression in some degree adequate to the truth of which I have been convinced— From a man in a waking moment—to men in their waking moments. Wandering toward the more distant boundaries of wider pastures— Nothing is so truly bounded & obedient to law as music—yet nothing so surely breaks all petty & narrow bonds.
>
> Whenever I hear any music I fear that I may have spoken tamely & within bounds. And I am convinced that I cannot exaggerate enough even to lay the foundation of a true expression— As for books & the adequateness of their statements to the truth—they are as the tower of Babel to the sky. (*PJ* 7:268–69)

This Journal entry finds its published version in the "Conclusion" chapter of *Walden* (324–25), but it is closely linked to the phenomenon of the sand foliage and to the plant-thinking that emerges from Thoreau's repeated observations and expressions. In the published text, Thoreau italicizes *"extra-vagant"* and *"Extra vagance!"* to emphasize the etymological idea of wandering "beyond the narrow limits of my daily experience" (324). The idea is already apparent in the Journal, even though it is connected to the narrow limits "of our ordinary insights & faith." Our general failing as ordinary thinkers becomes Thoreau's specifically literary fear in the "Conclusion."

In the Journal, literary language is so limited and narrow that books form "the tower of Babel," an apt image for a multitude of mutually unintelligible languages. The truth, on the other hand, is a unity, the sky. Thoreau seeks a language that will be "adequate to the truth," a point he repeats because he does not seem content with his expression. There may be no more than adequate expressions, but even such close approximations may express the truth of the writer's experiences and observations. In addition, the writer repeatedly translates written language into an image of speech, a man in a waking moment speaking to men in their waking moments. This image, in turn, seems to approach the vital language of the earth, the language that Thoreau calls, in the chapter "Sounds," "the language which all things and events speak without metaphor, which alone is copious and standard" (W 111).

The vital language of the earth speaks without metaphor and tells of constant transformation. It is not that the flowing railroad cut resembles leaves and lobes; rather, it is transformed into leaves and lobes, channels and arteries, strands and sands and banks and ripples. The very soil "fabulates" in a language that makes mythologies into mutterings, and that language is transformed into Thoreau's writings, both in the Journal and in *Walden*. If the writer's language prove itself adequate to the truth.

In at least three significant passages in "Spring" and the "Conclusion," Thoreau uses plant-thinking and botany to fabulate in an adequate language. Just after the sand foliage passage, he writes an elaborate paragraph detailing the first, early spring plants next to the withered vegetation of winter. The catalog of plants recalls earlier catalog passages from *Walden*, because the plants are common to the neighborhood and could be called "decent weeds, at least, which widowed Nature wears" (W 309). Thoreau's sincere affection for goldenrods, grasses, mulleins, and johnswort is apparent, and he recurs to the "architectural foliage" of acanthus, chicory, and other plants in describing "the arching and sheaf-like top of the wool-grass" as "among the forms which art loves to copy, and which, in the vegetable kingdom, have the same relation to types already in the mind of man that astronomy has. It is an antique style older than Greek or Egyptian" (310).

The ancient quality of common plants connects directly to the parable of the artist of Kouroo, whose timeless pursuit of perfection yields more than a staff (W 326–27). The salient motif is the wooden staff itself, which begins as a multitude of sticks rejected by the artist. The idea of the staff becomes "a stock in all respects suitable," a modest achievement. The stock of wood is transformed into "the proper shape," and the walking stick even becomes a writing instrument: "With the point of the stick he wrote the name of the last of that race in the sand, and then resumed his work" (327). Then the stick is smoothed and polished, becoming a proper "staff," and the artist finishes by putting on the "ferrule and the head adorned with precious stones." In the last transformation, the staff expands "into the fairest of all the creations of Brahma": "He had made a new system in making a staff, a world with full and fair proportions" (327). The transformations occur as extravagantly timeless efforts by the artist to create a practical, useful tool out of wood, a tool for traveling to unknown places. In addition to making more than a perfect staff—perhaps more even than a new system or world—the artist makes a perfect parable of artistry.

Last of all comes the old New England story of the "strong and beautiful bug which came out of the dry leaf of an old table of apple-tree wood" (*W* 333). The botanical language is understated but apt. The egg had been deposited in the living tree and had been "buried for ages under many concentric layers of woodenness in the dead dry life of society." Now, the bug emerges into a "beautiful and winged life," and the transformation awakens the narrator's strongest rhetorical question: "Who does not feel his faith in a resurrection and immortality strengthened by hearing of this?" (333). The story directly echoes a sentence from the sand foliage passage: "Thus, also, you pass from the lumpish grub in the earth to the airy and fluttering butterfly" (306). The resonating imagery of transformation is organic, whether it focuses on the artistic transformations of plants or the complete metamorphosis of insects from egg to beautiful, winged adult. For Thoreau, the stories tell one truth: the fundamental vitalism of the earth.

PART II

The Broken Task and
the Faithful Record

CHAPTER 4

Thoreau's Kalendar

Reading the Journal through Plants

Thoreau's Kalendar project is first inspired by plants and finally goes well beyond them. In all of its developments, the Kalendar is grounded in Thoreau's botany, and the botany is grounded in the Journal. The 1852 Journal features several lists of plants—their first leafing, their first blossoming, and their first fruiting. It features other seasonal occurrences, natural and human, so that the plants merge with personal events such as Thoreau wearing a light jacket for the first time in the autumn. The botanical lists of 1852 forecast the succession of Kalendar lists and charts that Thoreau develops over the final decade of his life. The Journal and the Kalendar project function together as faithful records of Thoreau's botanical excursions in the immediate region of Concord from 1852 to his death. They connect Thoreau and Concord to the seasonal world and make Concord into a more-than-local place, and they become openings for Thoreau, giving him new possibilities for recording his experiences and using the writings as prompts for further reflections. During his final years, the Journal records primarily his botanical explorations and observations: nearly every walk he takes is an excursion that features plants as a means of thinking—about his life, its purpose and meaning, and his work as a writer. Among possible functions, the Kalendar charts act as an index to the Journal, and they may map Thoreau's last decade in new ways.[1]

In Thoreau's last years, the Journal remains the main source and central focus of his writing. Reading across the decade, we find the development of Thoreau's botanical expertise coming together with his writing practices. One effect of the repeated practices of the botanical excursions is the production of knowledge: the immense number of recorded observations, identifications, descriptions, and experiences. The sheer volume of the Journal entries threatens to become overwhelming, both for the reader and for Thoreau. In his final years, the Kalendar project evolves out of the botanical lists of the Journal into separate lists and charts that summarize and order the Journal entries in monthly, seasonal, and annual formats.[2]

Important studies have been made of Thoreau's Kalendar project and phenology, but much remains to be discovered. Kristen Case's seminal 2014 article, "Knowing as Neighboring," focuses especially on the "General Phenomena" charts and employs the modern discipline of science studies to interpret Thoreau's Kalendar project as a phenomenological exploration of human / more-than-human relationships. Case has also launched a digital project of making the Kalendar charts available; as I write, many are still only available as manuscripts in the Morgan Library. The charts as monthly indexes provide a set of phenological observations that, as Case shows, are difficult to bring within standard literary interpretive frameworks.[3] Along with Robert D. Richardson's biography and Laura Dassow Walls's multiple studies, Daniel Peck's reading of the connection of the Kalendar to the Journal is balanced and persuasive.[4] The Kalendar serves as a large-scale phenological map, and for Thoreau it functions as a mnemonic guide to significant passages in the Journal. In addition, the Kalendar charts empower Thoreau to engage in the two major writing projects of his last years: *The Dispersion of Seeds* and *Wild Fruits*.

Along with the large charts in the Morgan Library, the Kalendar project includes many lists that serve as interim index material for the Journal. Several of these lists appear in manuscript form in the Huntington Library, the Houghton Library, and the Berg Collection of the New York Public Library.[5] By reading the lists in conjunction with the charts and the Journal passages to which they refer, we can begin to understand how Thoreau's writing process stems from his botanical excursions and from the plant-thinking he develops in the excursions. One hypothetical sequence of the Kalendar Project is as follows: first, journal entries and botanical studies develop rapidly to such a point (1852) that Thoreau creates summarizing botanical lists, focusing first on identifications and categories of plants in phenological terms; second, journal entries and botanical lists develop to such a point (1852 and after) that Thoreau creates other lists, called "Miscellaneous" and "General Phenomena" in the manuscripts, adding personal events and observations of fauna as well as added flora such as grasses, weeds, and cryptogams; and, third, Journal entries and lists in several categories develop to such a point (1860–62) that Thoreau creates the large-format monthly charts, making a summarizing verbal map of his botanical studies and excursions.

Within this framework, the most significant material remains the Journal entries themselves, read in their original context as well as in relation to the Kalendar lists and charts. One specific aspect of plant-thinking, seen in the Journal but not in the late charts, is Thoreau's reading of Asa Gray's *Manual of Botany*. Early on, Gray gives Thoreau the first inklings of the phenological work to be done. For example, the manuscript "Nature Notes: Flowers," in the Berg Collection of the New York Public Library, directly follows Gray, as Thoreau himself notes in his summary title: "The Flowering of Plants, accidentally observed in '51, with considerable care in '52; the Spring of '51 being 10 days, and more earlier than that of '52. The names those used by Gray. X observed in good season The XX before the names refer to '52 XXX very early—in—52."[6] The Journal of summer 1852 substantiates the central role Gray's nomenclature plays in Thoreau's work. Over the years, however, as Thoreau gains confidence in his skills of observation and identification, he engages Gray and his other botanical sources (Jacob Bigelow, George Emerson, F. A. Michaux, and even Linnaeus) in a scientific and experiential dialogue, one that always takes place in the pages of the Journal.

As the lists multiply, becoming records of annual events such as leafing, flowering, and fruiting, Thoreau engages in an experiential dialogue with himself. He seeks to identify larger patterns in his botanical excursions, multiplying the observations over succeeding years and beginning to frame the observations in seasonal interpretations. The Kalendar charts encompass hundreds of botanical excursions, placing them in monthly frames and presenting observations by date and year. In this sense, the Kalendar charts function as more than an index; they become an ordering device, a literary compass, a way of finding one's way through the maze of experiences and entries in the Journal. Even more than that, the Kalendar frame is also an opening. Like any map, a given chart could provide the means for developing strategies for exploration, expectation, and reflection. Perhaps as important, it could lead to new strategies for thinking and writing. And, for modern readers, the map is open to interpretation.

MAKING WHOLES OF PARTS

In the first days of 1852, Thoreau gives himself the outlines of the Kalendar project in terms of writing in the Journal. On 22 January, he tells himself to "set down such choice experiences that my own writings may inspire me.—and at last I may make wholes of parts" (*PJ* 4:277). The Journal instills the "habit of writing" and allows the writer to "remember our best hours—& stimulate ourselves" (4:277). As thoughts and memories proliferate, Thoreau imagines the cumulative effects of the entries: "Having by chance recorded a few disconnected thoughts and then brought them into juxtaposition—they suggest a whole new field in which it was possible to labor & to think. Thought begat thought" (4:277–78). At the outset of his botanical studies, Thoreau already sees how the excursions and the keeping of the Journal records will become a stimulating, mutually supportive process. He also sees how the parts could become wholes, a "whole new field" of work and thought.

It is certainly worth questioning whether plants are the parts that produce wholes, and whether the "whole new field" that Thoreau sees opening before him in 1852 is grounded in botany. The Journal supplies answers, both partial and complete. In an entry from 11 April, Thoreau observes *Alnus incana* plants blossoming and feels "a kind of resurrection of the year" (*PJ* 4:434). This common, widespread plant, speckled alder, is much like the skunk cabbage and its first flowering a week later. Both plants suggest to Thoreau that "this is the spring of the year" and prompt him to ask, "What occult relation is implied between this plant & man?" (4:467). As he works from the specific part toward a sense of a whole, he remarks that "for the first time I perceive this spring that the year is a circle— I see distinctly the spring arc thus far. It is drawn with a firm line. Every incident is a parable of the great teacher" (4:468). He asks deeper questions about "the mysterious relation between myself & these things" and, ever the surveyor, imagines making "a chart of our life—know how its shores trend—that butterflies reappear & when—know why just this circle of creatures completes the world" (4:468). Already in the spring of 1852, Thoreau is projecting the Kalendar charts he will actually create in 1860, and he is already taking a single phenomenon like the flowering of speckled alder or skunk cabbage to draw the firm line of a circle, one that encompasses plants, butterflies and other "creatures," and himself.

"At last I may make wholes of parts" serves as a kind of motto or epigraph for Thoreau's Kalendar project. It implies a long creative process that involves many steps and stages. The final decade bears ample witness to this process, and the Journal is a constant source for defining the ways in which plant-thinking plays a central role in Thoreau's creative imagination. Not least, however, is the daily work of identifying plants and singling out the parts themselves. In the botanical turn of 1851, Thoreau sets himself an immense task, and the Journal of 1852 to 1854 shows the steps he takes toward fulfilling it.

One of the most important steps is the making of lists. Thoreau begins listing identified species in 1851, but the practice becomes habitual in 1852, and it becomes a faithful practice in the years thereafter. In late April 1852, he makes a short list of *"early* willows in Mass according to Gray," one that begins as five species and expands to eleven by "putting [George B.] Emerson in brackets" (*PJ* 5:7).[7] In addition, Thoreau continually questions the relationships among the species and their botanists. On 14 May 1852, he notes, *"Hastily* reviewing this journal I find the flowers to have appeared in this order since the 28th of April—perhaps *some note* in my Journal has escaped me," and then follows this with a list of some thirty species (5:53–54). The most interesting aspect of this early work is Thoreau's self-deprecation, as if he feared someone might check his lists and find them less than accurate. It is also interesting that the list of flowers leads to a list of twenty-eight birds observed in the same period, all listed by common name (5:54–55), then to a truly miscellaneous list of early phenomena of the spring, including such facts as "frogs snore in the river" on 10 May (5:56).

As spring moves into summer, and the flowers multiply along with the near-daily excursions, the listings appear periodically, and Thoreau becomes more and more systematic. A remarkable but characteristic example comes in another *"hasty* review" and list on 24 June 1852. He lists "the flowers in the following order—I did not attend particularly to the trees, especially the evergreens—nor to the grasses &c &c. and have knowingly omitted several besides." He then lists 151 species over several pages (*PJ* 5:131–37), with dates given from 14 May to 23 June. He includes nine species of birds and "miscellaneous observations" that are a mix of botanical and zoological encounters (5:137–41). He makes efforts to be accurate and complete, so in many cases he adds species that he thinks he missed earlier. In the 25 July entry, for instance, he notes several flowers observed before 11 June that are still in blossom, several that have gone out of blossom, and several observed between 10 and 24 June that are still common (5:248–50).

A day later, he attempts to become yet more systematic in his summaries, developing a code for the flowers observed between 23 June and 27 July:

x   Those observed in very good season
xx   "   "   "   rather early
S   Those which have been in blossom for a day or two
X   "   "   some days
O   "   "   some time
V   not quite open (*PJ* 5:52)

The list follows the previous pattern, but birds are largely absent. Thoreau employs his phenological code for the list of some 130 species, all given by Latin name (5:252–56). "Miscellaneous observations within same dates" follows, summarizing observations noted in the Journal entries under those dates from 24 June to 25 July (5:257–59). For example, for 5 July he lists, "Season progresses to berrying time & locust weather" (5:258). In the 5 July Journal entry, he writes a paragraph that begins, "The progress of the season is indescribable" and ends, "We lie in the shade of locust trees—haymakers go by in a hay-rigging— I am reminded of berrying—" (5:185). The effect of these lists is that of summarizing observations, recording them by date, and keeping the writer's accounts straight. They function as interim summaries, a step toward the large-format charts of the Kalendar nearly ten years later.

While Thoreau continues to use parts of the phenological code to designate leafing and flowering in "good season," early, or late in the season, he does not adhere to the elaborate code as he has it in the summer of 1852. By May 1854, he is using the simple notations "X, XX, and XXX" to denote the three observed states. These also seem to function as phenological notes—that is, the annotated lists suggest Thoreau's developing expertise as a seasonal observer, noting whether or not he finds the leafing or blossoming of a plant as occurring in "good season." The developing expertise leads, in turn, to a proliferation of notes and observations, to plant-thinking extending in many directions. From 11 to 24 May 1854, for example, the Journal is full of notes on the leafing of trees and shrubs, and the phenological notations are incorporated directly into dated entries alongside other observations (*PJ* 8:113–48). In addition, this method of notation within the Journal entries allows Thoreau to move fluidly to other plants, as well as to observations of birds, other animals, and physical phenomena such as light and shadow. The Journal becomes a kind of ground-truthing exercise for the larger Kalendar project, if Thoreau was already intending to connect the many natural phenomena into one archetypal year, as David Richardson suggests. If a day can become a whole experience, perhaps a sequence of days could, too.

The entry of 24 May 1854 offers a fascinating summation of Thoreau's practices and the knowledge he gains from them. He gives succinct accounts of two excursions, one to Lee's Cliff in the very early morning, the other to Pedrick's Meadow in the afternoon. Both accounts describe the landscape in detail and supply a narrative to bring the reader into relation with the writer's experiences. In describing the afternoon excursion, Thoreau tells of wading into Beck Stow's swamp and finding the water "so cold at first that I thought it would not be prudent to stand long in it—but when I got further from the bank it was comparatively warm" (*PJ* 8:148). The patience leads to a discovery:

> Surprised to find the Andromeda polifolia in bloom & ap past its prime—at least a week or more— The calyculata almost completely done & the high blueberry getting thin. It is in water a foot & a half deep & rises but little above it— The water must have been several inches higher when it began to bloom— A timid botanist would never pluck it— Its flowers are more interesting than any of its family almost globular—crystalline white even the calyx except its tips tinged with red or rose—Properly called water andromeda— You must wade into water a foot or 2 deep to get it—the leaves are not so conspicuously handsome as in winter— (8:149)

The three plants Thoreau encounters in Beck Stow's swamp are common inhabitants of bogs and swamps in New England. All are closely related members of the *Ericaceae* or heath family: bog rosemary (*Andromeda polifolia*), leatherleaf (*Chamaedaphne calyculata;* identified as *Cassandra calyculata* in Gray, 317–18, and as *Andromeda calyculata* in Bigelow, 176–77), and the highbush blueberry (*Vaccinium corymbosum*). Thoreau is surprised to find the *Andromeda* in bloom, although it is past its prime. More interesting than the surprise is the botanist's persistence—even bravery—in wading into the water and plucking the flowers, so that his description of them can take on a salient specificity. While most of us would call the flowers pink, Thoreau's fine perceptions of color render them more particular and beautiful than one simple word can do.

The list of shrubs and trees that ends the 24 May entry of the Journal sums up the order of their leafing and includes "wild & a few tame" species (*PJ* 8:149). The plants are not listed by date, though some dates are given along the way. Instead of the X notations, Thoreau marks the list with a host of question marks and short remarks, as if he is not entirely confident in the dates of his observations. Many species he notes are "not seen," "not seen early," or "seen late," the observer seeming to take responsibility for any inaccuracy of dating (8:149–54). In the Princeton edition of the Journal, the editors provide "Later Revisions," Thoreau's ink and pencil revisions with dates for four succeeding years, 1855–58 (8:493–98). In addition, the "Historical Introduction" to volume 8 of the Journal includes a brief, excellent discussion about this material (8:399–400). Thoreau had begun listing the trees and shrubs in the order of their leafing in 1852, and 15 May 1853 has a list, too (6:125–26). Ultimately, the material from the years 1852 to 1858 culminates in a table called "Leafing of trees & shrubs in 52 . . . 60" (Howarth F27a); it features nearly 200 species and information on observations from 1852 to 1860 (8:400). These lists and revisions in volume 8 are close to the "Leafing of trees & shrubs" table. The "Later Revisions" material includes the Journal list (149–54), with the additions of dates and notes to show Thoreau's continued study. An entry like "Early trembles suddenly" (8:151), for example, is revised to read "Early trembles suddenly—Young May 1st—55 q.v. 2d May 4-56-1 inch over big as ninepence May 9–57 May 5–58" (8:495).

How, we might well ask, does this gathering of parts move toward a whole? And what form does such a whole take?

My account of Thoreau's process for compiling and organizing the parts of his world may be accurate, but it does not necessarily mean that the large charts of the Kalendar project are the only result of the successive lists and notations. A pair of entries in the March 1860 Journal suggest, in fact, a more discursive version of the writer's work. In the 22 March 1860 entry, Thoreau delineates the "phenomena of an average March" in a series of paragraphs (J XIII:208–10). He gives first the changing weather as winter turns toward spring, the frost comes out of the soil, and old leaves start drying in the woods. Then comes a paragraph on vegetation, in which the details become more prominent and salient. He names over twenty plants, wild and cultivated, and gives the dates that they bloom, or their sap begins to flow, or they put out catkins or spring buds and leaves. His old friend skunk cabbage appears, beginning to bloom on 23 March. He concludes that "one indigenous native flower blooms" (XIII:208–9). In a short paragraph, he summarizes the twenty-nine migratory birds that arrive, along with a few residents. Insects and worms and other small animals appear; mammals come forth; fish are seen floating in the streams and rivers; reptiles appear, especially turtles. He notes that the river opens on average by 5 March, a full eleven days before the Hudson. This trial summary in straightforward prose may well be an attempt at writing out the month in sentences. There is no narrative, no elaborate descriptive detail, no hint of the many excursions and Journal entries necessary for such a summation.

A second, more developed attempt comes on 25 March. After recounting an afternoon excursion to Well Meadow and Walden Pond, with observations of deer mice, a casual remark on common early spring plants like cress, sedge, skunk cabbage, senecio, *Viola pedata,* and *Caltha palustris,* along with the sighting of a pair of sparrows, Thoreau turns to "speak of the general phenomena of March" (*J* XIII:218). This second attempt reads like a draft introduction for a Kalendar essay. The first paragraph opens with the shifting weather of the month, "a tolerably calm, clear, sunny, spring-like day," and then remarks compassionately on the change from sleighing to "horses laboriously dragging wheeled vehicles through mud and water and slosh" (XIII:218). Over the course of ten well-wrought paragraphs, Thoreau describes the sleds, woodpiles, and snowplows lying abandoned in the March muck to deliver the wry theme in an epigram: "All things decay, / And so must our sleigh" (XIII:220).

After a near-digression contrasting walks in the winter and summer seasons on Colburn Hill, Thoreau proceeds with March, giving events by date and summarizing the kind of observations that mark the Journal and the Kalendar charts. This "Story of March," as the running title has it, runs nearly eight full pages in the 1906 edition of the Journal. The events are comprised of general phenomena and focus especially on the changing weather, on ice and snow and wind. The story begins with the event of spring some readers may call emblematic—the sand foliage: "Frost comes out of warm sand-banks exposed to the sun, and the sand flows down in the form of foliage" (*J* XIII:221). Some dates are given only cursory descriptions, while others are developed into marvelous short sketches:

> The 8th, it is clear again, but a very cold and blustering day, yet the wind is worse than the cold. You calculate your walk beforehand so as to take advantage of the shelter of hills and woods; a very slight elevation is often a perfect fence. If you must go forth facing the wind, bending to the blast, and sometimes scarcely making any progress, you study how you may return with it on your back. Perchance it is suddenly cold, water frozen in your

chamber, and plants even in the house; the strong draft consumes your fuel rapidly, though you have but little left. You have had no colder walk in the winter. So rapidly is the earth dried that this day or the next perhaps you see a cloud of dust blown over the fields in a sudden gust. (XIII:224)

This paragraph could stand as a Journal entry concerning an afternoon excursion in March. Each dated entry in the "Story of March" describes characteristic changes in temperature, wind, and weather. In this passage for the 8th, Thoreau combines the general and the specific in such a way as to render the experience in vivid details. The paragraph also brings the reader into the experience by repeating the "you" who has had no colder walk in the winter. In moving from the planning of the excursion to the walker's return to water frozen in your chamber at home, moreover, Thoreau takes you through an entire day. Parts become a dynamic whole within one paragraph.

The active effects of the excursion linger in many parts of the story, and they often combine the general and the specific. Take this description of an excursion on the 12th to a favorite destination: "Walden is melted on the edge on the northerly side. As I walk I am excited by the living dark-blue color of the open river and the meadow flood seen at a distance over the fields, contrasting with the tawny earth and the patches of snow" (J XIII:225). On the 16th, Thoreau launches his boat and makes "my first voyage for the year up or down the stream, on that element from which I have been debarred for three months and a half" (XIII:226). On the 17th, "when you go to walk in the afternoon, though the wind is gone down very much, you watch from some hilltop the light flashing across some waving white pines. The whole forest is waving like a feather in the wind" (XIII:227). A few days later, it is so warm that "perhaps I wear but one coat in my walk, or sweat in two. The genial warmth is the universal topic" (XIII:228). By the last day of the month, "the highways begin to be dusty, and even our minds; some of the dusty routine of summer even begins to invade them. A few heels of snow may yet be discovered, or even seen from the window" (XIII:229). For readers of the Journal, this summary narrative resonates strongly with the repeated March excursions from 1850 to 1861. The magic of the prose arises from the ways in which Thoreau records experiences in dynamic, concrete language.

The two entries from March 1860 suggest that Thoreau was drafting material for a monthly Kalendar, but it is not clear how lengthy each month might be or what elements it might contain. In the Journal entry of

26 March 1860, Thoreau follows his trial Kalendar stories with an excursion to Second Division Brook, and in some ways the writing evokes more vivid encounters than "general phenomena" call forth. As strongly descriptive as the trial material may be, Thoreau's response to direct experience leads to visual descriptions like this one: "The earliest willows are now in the gray, too advanced to be silvery,—mouse or maltese-cat color" (*J* XIII:232). Later in the same entry, however, he returns to summarizing and characterizing the month: "Tried by various tests, this season fluctuates more or less" (XIII:233). He details several phenomena as specific examples—whether or not there is sleighing; whether the river is open or closed; whether the air temperature is cold or warm; when the first skunk cabbage flowers (a range of some thirty-six days, by the way, 2 March 1860 to 6 April 1855 or 8 April 1854); when the first bluebird arrives; when the yellow-spotted tortoise can be seen; when the wood frog is heard (XIII:233–34). There is "a month's fluctuation, so that March may be said to have receded half-way into February or advanced half-way into April" (XIII:234). The trial narratives for March are testimonies to the fluctuations and to the season of change in Thoreau's almanac, and they testify as well to Thoreau's search for a prose style in which to deliver his experiences and insights.[8]

THE TRUE SAUNTERING OF THE EYE

The difficulties of making a whole out of parts are deeper than the straight-forward logistics, perhaps more overwhelming than the sheer amount of data to be sifted, organized, and rendered significant. One of the deepest problems Thoreau faces during the final decade is his own tendency to focus and study to excess. Already in an entry of 13 September 1852, he admonishes himself on this tendency:

> I must walk more with free senses—It is as bad to study stars & clouds as flowers & stones— I must let my senses wander as my thoughts—my eyes see without looking. Carlyle said that how to observe was to look—but I say that it is rather to see—& the more you look the less you will observe— I have the habit of attention to such excess that my senses get no rest—but suffer from a constant strain. Be not preoccupied with looking. Go not to the object let it come to you.
>
> When I have found myself ever looking down & confining my gaze to the flowers—I have thought it might be well to get into the habit of

observing the clouds as a corrective—But ha! that study would be just as bad— What I need is not to look at all—but a true sauntering of the eye. (*PJ* 5:343–44)

Even as he embarks on the botanical studies, excursions, and writing that will fill the next decade, Thoreau questions his method of intense observation, the "habit of attention" that plant-thinking fosters and develops. He recognizes that the attention to plants is not really the problem; instead, it is the way in which his attention confines his gaze to strict, limited objects. Any object, cloud as well as flower, can become an object of study, and that leads to a tyranny of the eye, the *looking* that involves a conscious seeking after knowledge and meaning. The opposite would be "a true sauntering of the eye," a phrase that also appears in "Walking," delivered as two lectures on 23 May 1852.[9] Such a freedom of seeing would be to "see without looking."

The tension between looking and seeing plagues Thoreau, even in his best months and days. In March 1853, for instance, he finds himself sick of "too much observing of nature" (*PJ* 6:30), and on 30 March 1853 he writes a nostalgic paean to his innocent past: "Ah those youthful days! Are they never to return? When the walker does not too curiously observe particulars, but sees, hears, scents, tastes, & feels only himself—the phenomena that showed themselves in him—his expanding body—his intellect & heart— No worm or insect quadruped or bird confined his view, but the unbounded universe was his. A bird is now become a mote in his eye" (6:56). Here the need for study and observation of parts seems somehow to have been eluded, for the phenomena of the world are already "in him—his expanding body—his intellect & heart." The whole is not confined by the eye, and the eye is not confined by the parts. Instead, the whole "walker" contains all the parts and makes a whole of them within himself. But when were these youthful days? Only a week later, in an entry of 7 April 1853, Thoreau swings to the other side, formulating a necessary process within the terms of the Kalendar project: "If you make the least correct observation of nature this year—you will have occasion to repeat it with illustrations the next, and the season & life itself is prolonged" (6:77). The fundamental process of repeating observations over successive years enriches the season and life itself, making both less fleeting and more "prolonged."

The opposition between the confinement of looking and the freedom of seeing is finally too stark, even facile. In an extended botanical excursion of

30 August 1856, Thoreau strikes a necessary balance between "a conscious and an unconscious life." The Journal entry is extensive and richly detailed, combining action and contemplation in even measures. Thoreau walks to "Vaccinium Oxycoccus Swamp," also known as Gowing's Swamp, in search of the "small cranberry, *Vaccinium Oxycoccus,* which [George] Emerson says is the common cranberry of the north of Europe" (*J* IX:35). The "small object" of the excursion is to compare the flavor of the small cranberry with the large or American cranberry, *Vaccinium macrocarpon* (Gray 314). This goal or "object" certainly invites the kind of preparation and observation that might lead to the tyranny of the eye, but instead the writer notes "the advantage of having some purpose, however small, to be accomplished,—of letting your deliberate wisdom and foresight in the house to some extent direct and control your steps" (IX:36). The advantage may indeed consist in the very slightness of the purpose, for that "business which is not your neighbors' business" actually "occupies territory, determines the future of individuals and states, drives Kansas out of your head, and actually and permanently occupies the only desirable and free Kansas against all border ruffians" (IX:36).[10] The exaggeration partly conceals Thoreau's concerns for bloody Kansas and the violence extending to Charles Sumner, the Massachusetts senator, and to John Brown; only the purpose of finding *Vaccinium Oxycoccus* allows him to drive Kansas out of his mind for an afternoon. Worse than either object, however, is not having an object or purpose at all, a "life . . . wholly without an object." For the writer as well as the walker, "both a conscious and an unconscious life are good. Neither is good exclusively, for both have the same source. The wisely conscious life springs out of an unconscious suggestion" (IX:37).

The suggestion, it seems, is that every life has expectations and seeks objects that can meet those expectations. In the story of the botanical excursion to Gowing's Swamp, Thoreau shows how the smallest of expectations and objects can expand to occupy large territories. We bear in mind that the object of the visit is to find and compare the small cranberry and common cranberry. Thoreau wades around the swamp for an hour, occasionally warming his feet on the surface of sphagnum moss (*J* IX:38–40). Observing and collecting specimens, the botanist notes that he is "the only person in the township who regards them or knows of them, and I do not regard them in the light of their pecuniary value" (IX:40). This "regard" is a special kind of looking and seeing, and it leads to more discoveries.

Thoreau finds an ant heap in the moss, then a new species of huckleberry, *Gaylussacia dumosa* var. *hirtella,* the latter giving him an opportunity to correct Gray's description (IX:41). The description of the "small black hairy or hispid berry" leads him to a new destination: "I seemed to have reached a new world, so wild a place that the very huckleberries grew hairy and were inedible. I feel as if I were in Rupert's Land, and a slight cool but agreeable shudder comes over me, as if equally far away from human society. What's the need of visiting far-off mountains and bogs, if a half-hour's walk will carry me into such wildness and novelty?" (IX:42). The new contrast, occasioned by the mix of expectation and surprise, is the discovery that the nearby is as wildly fruitful as the exotic or faraway. For several pages, Thoreau waxes eloquent on the topic, recurring constantly to the plants that have awakened him to the insight. Gowing's Swamp is no longer a local place, tamed by an owner's name. In the narrative of the botanical excursion, it is Vaccinium Oxycoccus Swamp, full of plants that "scarcely a citizen of Concord ever sees. It would be as novel to them to stand there as in a conservatory, or in Greenland" (IX:44). As the enthusiasm of his experience in the swamp grows, the writer's vision expands to encompass worlds: "These spots are meteoric, aerolitic, and such matter has in all ages been worshipped. Aye, when we are lifted out of the slime and film of our habitual life, we see the whole globe to be an aerolite, and reverence it as such, and make pilgrimages to it, far off as it is" (IX:45). The aerolite is a stony meteor, and the word also suggests a flight beyond the "slime and film of our habitual life." A botanical excursion to a swamp, with the conscious preparation and expectation of finding, observing, describing, and collecting a species of cranberry, follows unconscious suggestions, makes surprising discoveries, and reaches new worlds.

These processes are signal instances of plant-thinking, through which Thoreau balances the conscious and unconscious life. He engages in deliberate preparation and expectation, processes that can lead to a surprising discovery, a rich experience, or a deep reflection on the world and the work of writing. At its best, the botanical excursion leads to new, surprising worlds. Moreover, the writing of the excursion leads to such extraordinary places. A year later, to take one of many possible examples, in a passage from the 2 July 1857 Journal, Thoreau notes finding *Calla palustris* (Water arum) at Gowing's Swamp: "Having found this in one place, I now find it in another. Many an object is not seen, though it falls within the range of our visual ray, because it does not come within the range of our intellectual ray, *i.e.,* we are not looking for it. So, in the largest sense, we find only the world we look for" (*J* IX:466). This is, in the largest sense, the true sauntering of the eye. The achievement is such that Thoreau will return to the experiences and their effects in *The Dispersion of Seeds* and *Wild Fruits*.

Thoreau's perceptions of the world range through all the senses, but visual perceptions ground his botanical excursions and the entries in the Journal. He often remarks on birdsongs, to be sure, and works at identifying species by sound as well as by sight, and we know from his published works that he was immensely sensitive to sound. But his botany depends fundamentally on sight, with smell and taste filling secondary roles. The tension between looking and seeing, as well as the resolution of that tension in the true sauntering of the eye, suggests the central function of sight. Among many possible examples of this centrality is what Thoreau calls the "andromeda phenomenon," an exciting discovery he makes on an excursion of 17 April 1852. The ramble takes him to the three places he eventually names the Andromeda Ponds, between Walden and Fair Haven. He sees a large island patch of dwarf andromeda, which he follows Bigelow in identifying as *Andromeda calyculata*.[11] More important than the fact of identification is the vision Thoreau records. At first, he sees a "fine brownish-red color agreeable & memorable to behold" (*PJ* 4:461), and sees the sun reflected from the leaves, giving the whole a "greyish brown hue tinged with red" (4:462). The colors depend on his position directly opposite to the sun, and it is only there that he sees "this charming warm what I call *Indian* red color—the mellowest the ripest-red imbrowned color." When he climbs up the hill from the pond, he loses "that warm rich-red tinge—surpassing cathedral windows" (4:462). He tells himself to "look again at a different hour of the day & see if it is really so. It is a very interesting piece of magic. It is the autumnal tints in spring only more subdued & mellow" (4:462).

Two days later, on 19 April 1852, Thoreau returns to the pond and expands upon his sense of the significant phenomenon:

How sweet is the perception of a new natural fact!–suggesting what worlds remain to be unveiled. That phenomenon of the Andromeda seen against the sun cheers me exceedingly. When the phenomenon was not observed— It was not—at all. I think that no man ever takes an original or detects a principle without experiencing an inexpressible as quite infinite & sane pleasure which advertises him of the dignity of that truth he has perceived.

> The thing that pleases me most within these three days is the discovery of the andromeda phenomenon—It makes all those parts of the country where it grows more attractive & elysian to me. It is a natural magic. These little leaves are the stained windows in the cathedral of my world. At sight of any redness I am excited like a cow.—To-day you can find arrowheads for every stone is washed bright in the rain. (*PJ* 4:471)

The andromeda phenomenon is quintessentially visual, an effect of light perceived in a specific context and from a specific angle. It is not exactly a "natural fact" in Emerson's sense from the "Language" chapter of *Nature:* "Particular natural facts are symbols of particular spiritual facts" (20). True, the phenomenon transforms all the surrounding parts, making them more "attractive and elysian" than they were before, and the writer's world itself becomes a "cathedral." But even more important than these transformations is the natural fact that, like "stained windows," the andromeda suggest that "worlds remain to be unveiled" to the walker. The worlds are, moreover, ready to hand, even though most people will never encounter them or know they exist; they are like the *Azalea nudiflora,* awaiting discovery. In addition, the repeated experience recalls the ways in which one discovery can lead to another, the natural magic of an earthly vision moving toward more magic and other visions. As familiar as arrowheads are for Thoreau, for example, the repeated andromeda phenomenon lends them an air of excitement, "washed bright in the rain." The movement from one phenomenon to another also recalls the 1856 excursion to Vaccinium Oxycoccus Swamp, in which familiar and close-to-hand species become as exotic and magical as arctic plants. The preparation for one discovery can lead to the surprise of a second.

WRITING SEASONS

Laura Dassow Walls has recently edited *The Daily Henry David Thoreau,* a collection of quotations from the Journal and other works that shows how Thoreau "lived in season."[12] The source for the phrase is the Journal entry for 23 August 1853:

> Live in each season as it passes—breathe the air, drink the drink, taste the fruit, & resign yourself to the influences of each. Let them be your only diet drink & botanical medicines. In August live on berries, not dried meats &

pemmican as if you were on shipboard making your way through a waste ocean, or in a northern desert. . . . Why, "nature" is but another name for health, and the seasons are but different states of health. Some men think that they are not well in spring, or summer, or autumn, or winter; it is only because they are not *well in* them. (*PJ* 7:15)

The names for the four conventional seasons do not correspond with Thoreau's larger and more nuanced ideas about the seasons and how to live in them. August, for instance, is the season of berries, and to be healthy in that season is to live on berries, to take berries as your "botanical medicines." The seasons are always passing, rather than existing as stable, three-month entities. To "live in" the season means to pass as quickly as the season passes, embedding oneself in the passing of time. But the passing of time is in this sense not an abstraction, and certainly not the function of clocks or other measuring devices; instead, it is the concrete, physical changes of the natural world. For Thoreau, the seasons appear most fully in the changing forms and colors of the plants.

Like other forms of plant-thinking, a sense of the seasons requires sharp, repeated observations. Thus Thoreau remarks on this process in an entry from 27 October 1853: "Some less obvious and commonly unobserved signs of the progress of the seasons interests me most—like the loose dangling catkins of the hop horn-beam or of the black or yellow birch— I can recall distinctly to my mind the image of these things—& that time in which they flourished is glorious as if it were before the fall of man— I see all nature for the time under this aspect— These features are particularly prominent— As if the first object I saw on approaching this planet in the spring was the catkins of the hop-hornbeam on the hill sides—as I sailed by I saw the yellowish waving sprays—" (*PJ* 7:120). The second requirement for the sense of the seasons is memory. In writing the season of spring, Thoreau recalls the catkins of the hop-hornbeam and birches. These are not direct observations in late October, but rather the reminiscence of spring. Just the day before, in the entry for 26 October 1853, Thoreau had prepared for the catkins insight:

It is surprising how any reminiscence of a different season of the year affects us— When I meet with any such in my journal it affects me as poetry and I appreciate that other season and that particular phenomenon more than at the time.— You only need to make a faithful record of an average summer day's experience & summer mood—& read it in the winter—& it will carry you back to more than that Summer day alone could show—only the rarest flavor—the purest melody—of the season thus comes down to us. (7:115)

Memory is not always a function of writing, but in this passage Thoreau focuses on the "faithful record" that the writer makes and then reads "in the winter," so that only the very rarest flavor and purest melody of the season comes back to the reader. Contrary to my previous arguments, the passage does not highlight the visual aspect of such memories, though the entry of 27 October is a faithful record of "yellowish waving sprays." Here, taste and sound dominate—"the rarest flavor—the purest melody"—giving the memory a new and different sensory focus.

The strict requirements of repeated observations, faithful records, and acts of memory stem in part from the speed with which the seasons change. In an entry from 24 August 1852, Thoreau remarks that the year "is but a succession of days & I see that I could assign some office to each day—which summed up would be the history of the year" (PJ 5:313). The Kalendar lists and charts are already looming in such a passage. But the succession is diffi-cult to record faithfully because it passes so quickly: "Everything is done in season and there is no time to spare—The bird gets its brood hatched in season & is off. I looked into the nest where I saw a vireo feeding its young a few days ago—but it is empty—it is fledged & flown" (5:313). In this temporal frame, a season can be shorter than a few days or one day. It can occupy a moment, as in this passage from 11 August 1853: "What shall we name this season—This very late afternoon—or very early evening? This serene & placid season of the day most favorable for reflection—after the insufferable heats & the bustle of the day are over—& before the dampness & darkness of evening! The serene hour—the Muses' hour—the season of reflection" (6:298–99).

As the "season of reflection" suggests, Thoreau's style brings emotional tone to the perceptive sense of the seasons. The season of reflection is a brief transition between heat and bustle, on the one hand, and dampness and darkness, on the other. It is an emotional passage, a season within a season. As if the transitions of August arouse these reflections, Thoreau finds himself thinking and writing at length and depth on 18 August 1853:

What means this sense of lateness that so comes over one now—as if the rest of the year were down hill, & if we had not performed anything before—we should not now— The season of flowers or of promise may be said to be over & now is the season of fruits—but where is our fruit? The night of the year is approaching, what have we done with our talent? All nature prompts & reproves us— How early in the year it begins to be late. The sound of the crickets even in the spring makes our hearts beat with its awful reproof—while it encourages with its seasonable warning. It matters not by how little we have fallen behind—it seems irretrievably late. The year is full of warnings of its shortness—as is life— The sound of so many insects & the sight of so many flowers affect us so— The creak of the cricket & the sight of the Prunella & Autumnal dandelion. They say—for the night cometh in which no man may work. (*PJ* 6:306)

The last line echoes a biblical episode, in which Jesus heals the man born blind. Responding to the disciples' benighted queries, Jesus remarks that he has no time to judge the man or his parents. "I must work the works of him that sent me, while it is day: the night cometh, when no man can work. As long as I am in the world, I am the light of the world" (John 9:4–5). Unlike Jesus, Thoreau's sense of the season is that of a belated walker or writer, one who may be overtaken by the night. What work has the writer accomplished? Where is the fruit of his many labors? The belatedness weighs on the writer even in the season of spring, for even the season of "flowers or of promise" already carries its "seasonable warning." If the warning is in some ways an encouragement, it is also a reproof. It rings in the sound of the cricket and glows in the sight of the common plants *Prunella vulgaris* (heal-all) and fall dandelion (*Leontodon autumnalis*), signs that the night of the year approaches.

Not all seasons call forth such thoughts, and many times Thoreau is surprisingly hopeful. In an entry from 30 January 1854, for instance, he incorporates botanical imagery to encourage his thinking and writing: "The seasons were not made in vain— Because the fruits of the earth are already ripe—we are not to suppose that there is no fruit left for winter to ripen— It is for man the seasons and all their fruits exist. The winter was made to concentrate & harden & mature the kernel of his brain—to give tone & firmness & consistency to his thought— Then is the great harvest of the year—the harvest of thought— All previous harvests are stubble to this—mere fodder & green crop" (*PJ* 7:256). The definitions of "fruits," "ripeness," and "harvest" clearly expand beyond the literal, an expansion that Thoreau returns to and develops most fully in *Wild Fruits*. The most telling fruit, in this passage, is what we find in "the harvest of thought."

Even in the season of transition, Thoreau can focus on the hopeful imagery of harvest. In the entry of 7 August 1854, he sits writing in his attic and encourages himself in botanical figures: "Do you not feel the fruit of your spring & summer beginning to ripen, to harden its seed within you— Do not your thoughts begin to acquire consistency as well as flavor & ripeness— How can we expect a harvest of thought who have not had a seed time of character— Already some of my small thoughts—fruit of my spring life, are ripe, like the berries which feed the 1st broods of birds,—and other some are prematurely ripe & bright like the lower leaves of the herbs which have felt the summer's drought" (*PJ* 8:256–57). Two weeks later, still in the middle of creative drought, he recurs to the botanical excursions themselves as equally encouraging: "Walking," he writes in the entry for 22 August 1854, "may be a science—so far as the direction of a walk is concerned. I go again to the great meadows—to improve this remarkably dry season—& walk where in ordinary times I cannot go— There is no doubt a particular season of the year when each place may be visited with most profit & pleasure—and it may be worth the while to consider what that season is in each case" (8:288). Here, the encouraging thoughts are pragmatic, urging the walker to improve the season by learning where the season sends him.

Thoreau's seasonal imagination owes much to the botanical excursions, to phenological observations and their faithful recording in the Journal, and to the reflective entries that develop directly out of his botanical studies. The seasonal imagination, yet another important aspect of plant-thinking, finds concrete form in *The Dispersion of Seeds* and *Wild Fruits*, as I will argue in detail in following chapters. Moreover, it already takes compelling shape in "Autumnal Tints," the lecture/essay that becomes Thoreau's first post-humous publication in the pages of the *Atlantic Monthly*.

"AUTUMNAL TINTS" AND THE HARVEST OF THOUGHT

"Autumnal Tints" has its origins in the pages of the Journal, and in some ways it is the most complete, synecdochic, and visual version of the Kalendar project. From early in his botanical studies, Thoreau paid attention to the turning of colors in autumn leaves and to the fall of the leaves. As

Joseph Moldenhauer makes clear, the autumn of 1853 was a rich time for his interest, yielding forty Journal extracts used in the essay. The autumn of 1858 was even better, giving eighty Journal extracts used in the essay. As he worked on the piece in 1858, he was readying it for delivery as a lecture, which he gave four times between February and April 1859 and a final time (his last public lecture) on 11 December 1860, eight days after catching his death of cold on a botanical excursion.[13]

The most ambitious version of the project comes early on, in a Journal entry of 22 November 1853. Recording an excursion by boat, Thoreau interrupts his observations with a paragraph-long outline of the idea:

> I was just thinking it would be fine to get a specimen leaf from each changing tree & shrub & plant in Autumn in sep—& oct—when it had got its brightest characteristic color the intermediate ripeness in its transition from the green to the russet or brown state—outline & copy its color exactly with paint in a book—A book which should be a memorial of October—Be entitled October hues—or Autumnal tints— I remember especially the beautiful yellow of the P. Grandidentata & the tint of the scarlet maple. What a memento such a book would be—beginning with the earliest reddening of the leaves—woodbine & ivy—&c &c And the lake of rad—leaves—down to the latest oaks.
>
> I might get the impression of their veins & outlines in the summer—with lamp black—& after color them— (PJ 7:172)

This imagined book would be a kind of Kalendar of specimen leaves. The idea of the book is grandiose. It would require making rubbings, drawing outlines, and then painting each leaf individually and distinguishing the tints for each one. In "Autumnal Tints," Thoreau returns to the Journal and reproduces the entry faithfully:

> I formerly thought that it would be worth the while to get a specimen leaf from each changing tree, shrub and herbaceous plant, when it had acquired its brightest characteristic color, in its transition from the green to the brown state, outline it and copy its color exactly with paint in a book, which should be entitled October, or Autumnal Tints. Beginning with the earliest reddening-woodbine and the lake of radical leaves, and coming down through the maples, hickories and sumacs, and many beautifully freckled leaves less generally known, to the latest oaks and aspens. What a memento

such a book would be! You would need only to turn over its leaves to take a ramble through the Autumn woods whenever you pleased. Or if I could preserve the leaves themselves unfaded, it would be better still. I have made but little progress toward such a book, but I have endeavored instead to describe all these bright tints in the order in which they present themselves. The following are some extracts from my notes. (*Exc 225*)

The passage in "Autumnal Tints" registers the original ambition in all its details and then adds significant material of its own. The book that Thoreau envisions in 1853 is never to be published, and he admits that he has made "but little progress toward such a book." The idea of it, however, remains vivid. He even imagines a version that would include the actual leaves themselves, not merely painted drawings. In the essay, he adds the experience a reader would have, turning over the "leaves" of the book and thus taking a ramble through the autumn woods. The book would deliver illustrated botanical excursions, taken any time the reader might please. In the *Atlantic Monthly*, "Autumnal Tints" reproduces only one such specimen leaf, the famous image of a scarlet oak, and it is not reproduced in color. Still, Thoreau makes a set of three implicit promises: his descriptions will evoke the bright tints of autumn; he will present the colors in the chronological order of a botanical Kalendar; and his Journal will provide the accounts of his excursions.

The Kalendar project looms large in this set of promises. The order of presentation follows the phenological order of autumn leaves turning color. The internal headings outline the order, beginning with "The Purple Grasses" around 20 August. "The Red Maple" begins by 25 September; "The Elm," about 1 October or later; "Fallen Leaves," 6 October; "The Sugar Maple," by 17 October; "The Scarlet Oak," by 26 October. The lecture/ essay thus presents a discussion in seasonal arrangement, and this seasonal aspect delivers the experience of encountering the autumnal tints as they appear. It is as if the reader or listener were following Thoreau on a series of botanical excursions, in some ways like the experience of reading the "Story of March" in the Journal of March 1860. The accounts are characteristic, a summation of multiple autumns and many fallen leaves. Even though the headings begin with precise dates, the actual discussions are much more given to a range of experiences, so that a particular species is shown to exhibit its colors over a range of dates. This suggests, in a wholly interesting way, that Thoreau is giving a characteristic seasonal order of excursions and, at the same time, evoking the multiplicity of years and excursions that have contributed to making this version of the Kalendar.

The hybrid status of "Autumnal Tints" as lecture and essay should be emphasized, for it calls for a dynamic interpretive model of the piece. On the one hand, it is important that Thoreau writes and delivers the lecture four times in February–April 1859. The lecture comes immediately after the field work of the fall of 1858, and the Journal accounts of August–November 1858 are the most numerous and significant in the writing. On the other hand, it is certainly notable that this Kalendar work comes two full years before Thoreau catches his cold in the Concord woods on 3 December 1860. When he delivers the lecture a final time on 11 December 1860, Thoreau is bidding farewell to the grand project of the Kalendar and the harvest of the "Autumnal Tints" book he had envisioned.

The 1853 Journal already records the creative significance of the harvest. In the early days of October, most of Thoreau's entries are quite short, but the onset of autumn colors and the fall of leaves causes him to write at length. After a week of warm weather, like a second summer, he writes on a rainy 22 October, "I cannot easily dismiss the subject of the fallen leaves—how densely they cover and conceal the water for several feet in width under & amid the alders and button bushes & maples along the shore of the river—still light tight & dry boats—dense cities of boats—their fibres not relaxed by the waters—undulating & rustling with every wave— Of such various pure & delicate though fading tints—of hues that might make the fame of teas—dried on great nature's coppers" (PJ 7:105–6). The fall of the leaves is entangled with the "pure & delicate though fading tints," an entwining of death and beauty. The leaves multiply, becoming "dense cities of boats," but the undercurrent is darkly present.

The persistent image of death, as foundational as it is in Thoreau's plant-thinking, is harvested by the writer's faithful imagination:

Consider what a vast crop is thus annually shed upon the earth— This more than any mere grain or seed is the great harvest of the year— This annual decay and death—this dying by inches before the whole tree at last lies down & turns to soil. As trees shed their leaves—so deer their horns & men their hair or nails. The years great crop—I am more interested in it, than in the English grass alone or in the corn. It prepares the virgin mould for future corn fields—On which the earth fattens— They teach us how to die. How many flutterings before they rest quietly in their graves— A myriad wrappers for germinating seeds— By what subtle chemistry they will mount up again—climbing by the sap in the trees. The ground is all particolored with them. For beautiful variety can any crop be compared with them! (*PJ* 7:106–7)

In a way reminiscent of Walt Whitman's 1856 poem, "This Compost," the imagination transforms death through language. Repeating terms like "vast crop," "great harvest," and "great crop," Thoreau joins the agricultural imagery to the observation that leaf mold protects germinating seeds of a future forest. The "subtle chemistry" is in fact the botanical transformation of seeds to trees and other plants, and the real "crop" of the fall is the "beautiful variety" of fallen leaves and the regenerative, vital power they embody in their flutterings. In that season, the leaves and their autumnal tints teach us how to die.

These passages from the 22 October 1853 Journal appear in expanded form in "Autumnal Tints," joined with entries from 20 October and 23 August 1853. In a series of eight paragraphs, ending the "Fallen Leaves" section (*Exc* 240–42), Thoreau establishes a narrative frame of a boat excursion on 16 October and recurs to the imagery of the "great harvest of the year," the beautiful variety of the crop, and the flutterings that precede the sinking down. So "they teach us how to die." By rearranging sentences and gathering images in the published essay, he accentuates the active, transformative qualities of the leaves. The ground may be all particolored with the dead leaves, but "they still live in the soil whose fertility and bulk they increase." In the rush of sound and action, the leaves rustle and whisper, rise and fall, until they are "transmuted at last," to become rising sap. And the sapling sheds its first fruits in becoming "the monarch of the forest" (241). If we are taught how to die, we learn also to regard the "cemetery of the leaves" as a proper place for a pleasant, truth-filled walk.

The closing paragraph returns us to the action of an excursion, but now "the whole earth is a cemetery pleasant to walk in." The paragraph has a partial source in a Journal entry of 29 October 1855, but the final sentences revise and expand the original: "What though you own no lot at Mount Auburn? Your lot is surely cast somewhere in this vast cemetery, which has been consecrated from of old. You need attend no auction to secure a place. There is room enough here. The Loose-strife shall bloom and the Huckleberry-bird sing over your bones. The woodman and hunter shall be your sextons, and the children shall tread upon the borders as much as they will. Let us walk in the cemetery of the leaves,—this is your true Greenwood Cemetery" (*Exc* 242). The earnest humor of the address surely betrays a trace of the lecture hall. The "lot" Thoreau offers is common to us all, and the "vast cemetery" offers more room than the two celebrated, consecrated burying grounds in Cambridge and Brooklyn. Nor is the direct address critical of the audience; instead, the speaker invites us all, "Let us walk in the cemetery of the leaves." Your true green wood.

As important as these early Journal entries are, "Autumnal Tints" harvests most of its Journal extracts from the fall seasons of 1857 and 1858.[14] The late entries are too numerous to discuss in detail, even here, but they contribute significantly to Thoreau's meditation on "the intentions of the eye," the theme that drives the "Scarlet Oak" and concluding sections of the essay. The visual beauty of the "immense harvest which we do not eat" (*Exc* 224) is clear from the outset, but in "The Scarlet Oak" Thoreau celebrates both the form and the colors of the leaves. The illustration accompanying the text, reproduced from the *Atlantic Monthly* publication, embellishes the rough sketch in the 11 November 1858 entry of the Journal (*J* XI:314), while the brief, enthusiastic praise of the form in the Journal entry becomes five well-developed paragraphs (*Exc* 249–52). The color of the scarlet oak varies according to place and light, and, in that regard, it resonates with the andromeda phenomenon. Thoreau describes a particular view from "a cliff in the southwest part of our town" (certainly Lee's Cliff), with the sun getting low in a late October sky and the woods in Lincoln, "south and east of me . . . lit up by its more level rays" (253). On such a characteristic excursion, the scarlet oaks display "a more brilliant redness than I had believed was in them." Seen at a distance, every tree becomes "a nucleus of red, as it were, where, with the declining sun, that color grows and glows. It is partly borrowed fire, gathering strength from the sun on its way to your eye" (254).

A Journal entry from 2 November 1858 picks up the perspective on a dark day, and the botanical excursion that day gives a surprisingly bright view of the "deep, dark scarlet and red, the intensest of colors, the ripest fruit of the year" (*J* XI:277). Thoreau sees November as the "twilight of the year" (XI:273), but this only improves the function of the eye. The botanical excursions run through the lecture / essay from the beginning, re-enacting the studies that yield the Kalendar project. As he nears the end of the day, the year, and the performance, Thoreau turns to address his audience directly:

> Let your walks now be a little more adventurous; ascend the hills. If, about
> the last of October, you ascend any hill in the outskirts of our town, and
> probably of yours, and look over the forest, you may see—well, what I have
> endeavored to describe. All this you surely *will* see, and much more, if you
> are prepared to see it,—if you *look* for it. . . . Objects are concealed from
> our view, not so much because they are out of the course of our visual ray
> as because we do not bring our minds and eyes to bear on them; for there
> is no power to see in the eye itself, any more than in any other jelly. We do
> not realize how far and widely, or how near and narrowly, we are to look.
> The greater part of the phenomena of Nature are for this reason concealed
> from us all our lives. (*Exc* 256)

No hint here of a tension between looking too hard and seeing too little.
No wistful yearning for the true sauntering of the eye. The Journal source
for this passage comes in the entry of 4 November 1858, although he narrates a climb at the last of October. In the Journal, he writes, "Objects are
concealed from our view not so much because they are out of the course
of our visual ray (continued) as because there is no intention of the mind
and eye toward them" (*J* XI:285). This double intention evokes the process
of preparation, study, and expectation that can lead to discovery.

In exhorting his audience, Thoreau joins seeing and looking in the act of
preparing both our minds and eyes. In a sense, he argues, the leaves of the
scarlet oak must "be in your eye when you go forth":

> We cannot see anything until we are possessed with the idea of it, take it
> into our heads,—and then we can hardly see anything else. In my botan-
> ical rambles, I find that, first, the idea, or image, of a plant occupies my
> thoughts, though it may seem very foreign to this locality—no nearer than
> Hudson's Bay—and for some weeks or months I go thinking of it, and ex-
> pecting it, unconsciously, and at length I surely see it. This is the history of
> my finding a score or more of rare plants, which I could name. (*Exc* 257)

The Journal entry for 4 November 1858 finishes this process of how a plant
occupies Thoreau's thoughts and comes into his eye with a remarkable
phrase: "at length I surely see it, and it is henceforth an actual neighbor of
mine" (*J* XI:285–86). The phrase recalls the chapter "Brute Neighbors" from
Walden as well as Case's essay on the Kalendar, "Knowing as Neighboring,"
and it suggests a way in which Thoreau intends the Kalendar project to teach

a new "intention of the mind and eye." Moreover, in this specific process of preparation, expectation, looking, seeing, and neighboring, it establishes a sequence of plant-thinking in all its richness and variety. The sequence may be better described as a constellation of practices, moving in numerous possible directions. Such a constellation, always evolving, could enact the kind of free sauntering experienced by the most deeply rooted of plants.

The composition of "Autumnal Tints" constantly evokes the dynamic practices of plant-thinking. The mosaic of Journal entries displays the ways in which Thoreau locates and collocates passages for greatest effect. The "intention of the mind and eye" recurs to the phrases "intention of the eye" and "different intentions of the eye and of the mind," used in the Journal entry of 9 September 1858 (J XI:153). In the lecture/essay, the paragraph from the 9 September entry comes within the sequence of three long paragraphs from the 4 November 1858 Journal (XI:284–87), but it fits neatly into the four-paragraph sequence devoted to the preparation, expectation, looking, seeing, and neighboring:

> A man sees only what concerns him. A botanist absorbed in the study of grasses, does not distinguish the grandest pasture oaks. He, as it were, tramples down oaks unwittingly in his walk, or at most sees only their shadows. I have found that it required a different intention of the eye, in the same locality, to see different plants, even when they were closely allied,—as *Juncaceae* and *Gramineae;*—when I was looking for the former, I did not see the latter, in their midst. How much more then it requires different intentions of the eye and of the mind, to attend to different departments of knowledge! How differently the poet and the naturalist look at objects! (*Exc* 257)

The salient point of this passage is the way in which the focused idea of intention broadens to take in different intentions of the eye and mind required in different departments of knowledge. The botanist may well seek to see not only the grasses but also the rushes. So, too, the poet and naturalist look at plants differently, and Thoreau partakes of both intentions of the eye and mind. His botany is neither science nor poetry, but a constellation of intentions and practices.

We can call that constellation the Kalendar project, or Thoreau's botany, or plant-thinking. Whatever we call it, the final sentence of "Autumnal Tints" evokes its peculiar, faithful combination of humility and ambition:

"When you come to observe faithfully the changes of each humblest plant, you find that each has sooner or later its peculiar autumnal tint, and if you undertake to make a complete list of the bright tints it will be nearly as long as a catalogue of the plants in your vicinity" (*Exc* 259). Nuanced and provisional, this statement has its source in the Journal entry of 22 October 1858, in which Thoreau notes that he believes "*all* leaves . . . acquire brighter colors just before their fall" (XI:240). Some writers, too. For Thoreau faithfully keeps to the project of observations and excursions, intentions and practices, that leads him through his last bright seasons.

CHAPTER 5

The Dispersion of Seeds and the Writer's Faithful Record

On 20 September 1860, Thoreau made his "Address on the Succession of Forest Trees" at the annual Middlesex County Cattle Show in Concord. Nine days later, he sent the manuscript to Horace Greeley, hoping he might publish this "part of a chapter on the Dispersion of Seeds."[1] The main passages that form "Succession of Forest Trees" appear in various parts of the manuscript "Dispersion of Seeds," edited under the title *Faith in a Seed* by Bradley Dean.[2] As the editors of *Excursions* explain, Thoreau worked on the larger project of *The Dispersion of Seeds* for several months after delivering the "Succession of Forest Trees" address and publishing it in Greeley's *New-York Weekly Tribune*. The subject was of abiding interest to the writer, for the study of plants engaged his fundamental ideas of faith, temporality, and nature as a vital force, or *phusis*.

In *Seeing New Worlds,* Laura Dassow Walls argues that "Succession of Forest Trees" and *The Dispersion of Seeds* are fundamentally about the idea of succession and the Darwinian theory of evolution through natural selection.[3] A related frame is that of geographical distribution and biogeography. Like Darwin in *Voyage of the Beagle,* Thoreau is thinking through the spatial and temporal dimensions of plants and other organisms in the landscape. The ecological process of succession provides the means for developing themes of spatial dispersion and temporal succession. Together, dispersion and succession explain how landscapes change, how one species of forest tree replaces another. These patterns of succession are in turn affected by human agency, by the landowners and their practices of cutting, burning, planting crops, and allowing the forests to regrow. In addition to these external patterns, Thoreau's faith in his own work lies in *phusis*, the in-dwelling dynamism of the earth. This further suggests that the writer's faithful record of earthly succession records his own development as a writer.

The second aspect of succession evokes the process of writing. As I argued in part 1, Thoreau's published works themselves are records of successive stages of writing. *The Maine Woods, Cape Cod,* and *Walden* accord with the idea of the Journal as a faithful record of Thoreau's ways of knowing, rather than as a statement of knowledge or as a workshop for drafting and revising essays, lectures, and books. This is to interpret Thoreau's practice of writing as itself both successive and dispersive. In addition, for Thoreau the repeated botanical excursions become a model for ways of knowing through writing. These overlapping aspects of writing suggest that the narrative model of the botanical excursion can help us read *Dispersion of Seeds* as a faithful record on multiple levels.

Thoreau's faithful record of observations and excursions relating to plant succession reaches as far back as July 1850 (*PJ* 3:92), and the text of "Succession of Forest Trees" features material from scattered entries for 1852–60 (*Exc* 544). By contrast, the Journal of October–November 1860 provides the most concentrated material for *The Dispersion of Seeds*. It is as if the process of writing and presenting "Succession of Forest Trees" at the cattle show spurs Thoreau to an even more intense study of dispersion and succession. In fact, the Journal entries for the two months in 1860 are faithful records of his persistent, focused attention to the woodlots of the Concord area. Thoreau visits dozens of woodlots repeatedly, making measurements, collecting samples, making and analyzing observations, and drawing inferences from the concrete data. The process resembles that of a field ecologist. He is no longer simply listing species and identifying them by their constellation of key characteristics or by their phenological appearances. Rather, he is finding patterns in the stands of trees and the seedlings growing beneath them; he is counting annual growth-rings on decaying stumps, on saplings, sprouts, and seedlings. The Journal of October–November 1860 records these new, expanded insights for *The Dispersion of Seeds*, both in content and in inspiration.

Bradley Dean's title for his edition of the *Dispersion of Seeds* manuscript, *Faith in a Seed*, registers Thoreau's perspective about his own botanical studies and their larger importance. Interestingly, the phrase does not appear in *The Dispersion of Seeds*. Instead, we find it in a paragraph near the end of "Succession of Forest Trees," as Thoreau is winding up his arguments about the unacknowledged role of dispersion and succession in the Concord neighborhood: "Though I do not believe that a plant will spring up where no seed has been, I have great faith in a seed—a, to me, equally mysterious origin for it. Convince me that you have a seed there, and I am prepared to expect wonders. I shall even believe that the millennium is at hand, and that the reign of justice is about to commence, when the Patent Office, or Government, begins to distribute, and the people to plant the seeds of these things" (*Exc* 181–82). This passage is characteristic of Thoreau's humor, always with a serious bite to it. The seeds that yield the "millennium" and "the reign of justice" recall "The Bean-Field" in *Walden* and "such seeds . . . as sincerity, truth, simplicity, faith, innocence, and the like" (*W* 164). In "Succession," however, the actual seeds give rise to the speaker's faith in the dynamic processes of the earthly world. This faith may be related, in turn, to a passage in the important 16 October 1860 entry of the Journal that does not make it into *The Dispersion of Seeds,* either, but resonates strongly with Thoreau's earthly, botanical faith: "As time elapses, and the resources from which our forests have been supplied fail, we shall of necessity be more and more convinced of the significance of the seed" (*J* XIV:132).

As Thoreau was drafting the three chapters of *The Dispersion of Seeds,* he was drawing from the Journal entries and his experiences on repeated botanical excursions to forge his ideas into a coherent, book-length essay that would demonstrate "the significance of the seed" as well as his "faith in a seed." At the same time, we recognize that *The Dispersion of Seeds* is an incomplete draft, though it gives clear lineaments of the book that would have emerged from revisions. Moreover, in *The Dispersion of Seeds* the excursions and Journal entries show abundantly how Thoreau employs his narrative voice to call attention to the dates and circumstances of his data-gathering. He transposes paragraphs and entire pages from the Journal to *The Dispersion of Seeds,* but any given page can also feature material from widely separated dates, creating a kind of temporal mosaic of writing. In addition, the structure of the book is neither chronological nor seasonal, and in that way it is a departure from Thoreau's other published works. Instead, *The Dispersion of Seeds* organizes the botanical excursions recorded in the Journal into a discursive form, dispersing excursion material according to topic. At strategic moments, Thoreau incorporates his scholarship and reading into the discussions. The primary impetus throughout *The Dispersion of Seeds,* however, is the writer's sense of discovery, which builds within each chapter and from chapter to chapter, creating an effect of active learning.

The text of "Succession of Forest Trees" plays a significant role in chapter 1 and chapter 2 of *The Dispersion of Seeds*. Thoreau himself sees "Succession of Forest Trees" as a key part of the larger project. For example, "Succession" provides two important paragraphs in the introduction to *The Dispersion of Seeds*:

> When, hereabouts, a forest springs up naturally where none of its kind grew before, I do not hesitate to say that it came from seeds. Of the various ways by which trees are known to be propagated—by transplanting, cuttings, and the like—this is the only supposable one under these circumstances. No such forest has ever been known to spring from anything else. If anyone asserts that it sprang from something else, or from nothing, the burden of proof lies with him.
>
> It remains, then, only to show how the seed is transported from where it grows to where it is planted. This is done chiefly by the agency of the wind, water, and animals. The lighter seeds, as those of pines and maples, are transported chiefly by wind and water; the heavier, as acorns and nuts, by animals. (*Faith* 24)

With a few minor differences, these two paragraphs are identical to the passage in "Succession of Forest Trees" (*Exc* 166–67). As in the passage regarding "faith in a seed," Thoreau implicitly rejects the notion of spontaneous generation, an idea that had broad currency among followers of Louis Agassiz in the United States. Plants spring from seeds—an axiomatic fact. The question of how a forest can change from one species to another becomes a question of how the seeds of the succeeding plants are transported and planted, and how they become the dominant trees in a woodlot. "Succession of Forest Trees" is a key to stating the problem Thoreau addresses; *The Dispersion of Seeds* records the well-developed answers he discovers.

FROM SO SIMPLE A BEGINNING

Chapter 1 of *The Dispersion of Seeds* (*Faith* 23–104), the longest of the three, treats the transportation of seeds from plants to plantings. Thoreau presents the agency of wind, water, and animals by categorizing plant species according to their reproductive strategy of dispersion. The first class of species, for example, includes the "light and winged seeds of trees" and focuses on the agency of wind (*Faith* 67). The order of species seems to follow the importance and abundance of the forests. In a Journal entry of 17 October 1860, Thoreau lists over a dozen species of trees by these criteria, first numbering white and pitch pine; oaks; white birch; red maple; chestnut; and hickory. Then, without specific number, he lists another eleven common trees: alder; hemlock, spruce, and larch; cedar (white and red); willow; locust; apple; red cherry; sugar maple (*J* XIV:134). Chapter 1 does not follow this order exactly, but it comes quite close, and the first seven trees are clearly the most important and abundant in the Concord woods.

The pitch pine comes first and is treated in detail. Even though wind is the main method of seed dispersion, squirrels come in for extended description and discussion, based on the Journal entries of October 1860. Then follow white pines, hemlock and larch, birches, alder, maples, elms, ash, willows and poplars, and the buttonwood, as Thoreau calls the sycamore (*Faith* 24–67). The pitch pine and white pine constitute the longest section (24–39), and in some ways the pine woodlots provide the framework for all of Thoreau's ideas about the dispersion of seeds and succession of forest trees. The pitch pine is so common in New England that "all my readers probably are acquainted with its rigid conical fruit, scarcely to be plucked without a knife—so hard and short that it is a pretty good substitute for a stone" (24). The cone is a figure of simplicity and persistence, for cones can hang on a tree for years, and they often persist through the winter. The pitch pine is exceptionally productive of seeds, especially in rocky, sandy, or poor soil, and the seeds can be dispersed all winter, blowing over snow to come to rest and "rise up pines" (27). In addition to the wind, squirrels are vital to the dispersion. Thoreau describes how they gnaw off the twigs and carry away the cones to be plucked and stripped. He admires the dexterity and cleanliness of the squirrel's work, showing how perfectly matched are seed and planter (30–32).

According to Thoreau, red squirrels are associated with white pines, and he asserts that anyone collecting white pine seeds is indebted to the red squirrel's labor (*Faith* 37). By cutting off cones and burying them, they provide a repository of seeds, even though white pines produce seeds sparingly. Thoreau recounts a humorous episode from his 9 September 1857 Journal in which he tries to compete with the squirrels for the seeds. He climbs trees well enough, but then he finds that the cones "are now all flowing with pitch, and my hands are soon so covered with it that I cannot easily cast down my booty when I would, it sticks to my fingers so; and when I get down at last and have picked them up, I cannot touch my basket with such hands but carry it on my arm, nor can I pick up my coat which I have taken off unless with my teeth. . . . How fast I could collect cones if I could only contract with a family of squirrels to cut them off for me!" (39). While the large lessons of nature's perseverance and steady progress come through the treatment of the white pine and its dispersion strategies (36–37), Thoreau is equally interested in providing a good story and a sense of joyful experience. The botanical excursion of 9 September 1857 delivers both.

As the examples of the squirrels suggest, the length and detail of Thoreau's discussions do not depend only on the importance and abundance of the species, but also on the writer's appreciative interest and his eye for a good story. For instance, the willows and poplars garner extended treatment (*Faith* 55–65), not because of their importance so much as because their downy seeds illustrate the fecundity, hardiness, and variety of common woody plants. Willows and poplars seem to spring up anywhere and everywhere:

> No soil is so dry and sandy, none so wet, scarcely any so alpine or arctic, but it is the peculiar habitat of some species of willow. When I was at the White Mountains in July 1858, considerable tracts of its alpine region were hoary with the down of the little bearberry willow (*Salix uva-ursi*), a densely tufted trailing shrub on which you trod as on moss. Its seeds were just bursting away with irrepressible elasticity and buoyancy, and spreading its kind from peak to peak along the White Mountain range. Another also found there, *Salix herbacea,* the smallest of all willows if not of all shrubs, is said, together with the *Salix arctica,* to extend the furthest northward of all woody plants. (58–59)

In addition to hardiness and flexibility, willows illustrate the immense variety of plant forms and the invisible abundance of their seeds, and for Thoreau they are also a lesson in humility: "Ah willow, willow, would that I always possessed thy good spirits; would that I were as tenacious of life, as *withy,* as quick to get over my hurts" (63–64). The source for this reflection appears in the 18 March 1861 entry in the Journal. At the time, Thoreau was seriously ill; the previous December, he had caught the chest cold that would eventually kill him. But, in the Journal, as in this passage of *The Dispersion of Seeds,* he is in fact like the willow—all buoyancy and good spirits, or at least he aspires to be (*J* XIV:328–29).

The extensive first section on winged seeds ends with a five-paragraph meditation on the vital power of plants (*Faith* 66–67). The governing idea is set in the first sentence: "From such small beginnings—a mere grain of dust, as it were—do mighty trees take their rise" (66). The sentence may echo the proverb "Little strokes fall great oaks," a truism that Thoreau reverses in an earlier passage: "They are few who consider what little strokes, of a different kind and often repeated, *raise* great oaks or pines. Scarcely a traveller hears these or turns aside to communicate with that Nature which is steadily dealing them" (37). The sentence may allude to William Bradford's *Of Plymouth Plantation* ("Thus out of small beginnings greater things have been produced by His hand"), a book Thoreau knew very well. To my hearing, however, it strongly echoes the famous last sentence of Darwin's *Origin of Species*: "There is a grandeur in this view of life, with its several powers, having been originally breathed into a few forms or into one; and that, whilst this planet has gone on cycling according to the fixed law of gravity, from so simple a beginning endless forms most beautiful and most wonderful have been, and are being, evolved."[4] The parallel of Darwin and Thoreau emphasizes a cosmic importance in the "mere grain of dust," a beginning that, simple and small, gives rise to mighty trees and other endless forms. Thoreau directly cites Pliny and John Evelyn to support the idea of a minute seed becoming a tree of "gigantic dimensions" (*Faith* 66), then wonders what those early natural historians would have said if they had seen the giant sequoia of California, springing from such a small seed and yet outlasting many of the earth's kingdoms (67). In response to his own question, he imagines the earth itself springing from a seed, not giving rise to the giant sequoia but to a common willow tree. By this figurative ratio, such a "seed of the earth" would be a "globe less than two and a half miles in diameter, which might lie on about one-tenth of the surface of this town" (67).

Thoreau encountered and studied Darwin's *Origin of Species* in the early weeks of 1860, an experience that his biographers have richly described and analyzed.[5] In addition to specific references to Darwin's botanical studies, Thoreau draws on the "development theory" of *Origin of Species* to emphasize the generative power of nature, especially in plants. In the section on white pines, for example, he notes how these trees easily go unnoticed until they seem to spring up in a handsome white pine woodlot. The solution to this apparent puzzle is not spontaneous generation but the steady perseverance of seed production and dissemination:

> We need not be surprised at these results when we consider how persevering Nature is and how much time she has to work in. It does not imply any remarkable rapidity or success in her operations. A great pine wood may drop many millions of seeds in one year, but if only half a dozen of them are conveyed a quarter of a mile and lodge against some fence, and only one of these comes up and grows there, in the course of fifteen or twenty years there will be fifteen or twenty young trees there, and they will begin to make a show and betray their origin. (*Faith* 36)

The source for this passage is the Journal entry for 7 November 1860. In the excursion of that afternoon, Thoreau notices a number of little white pines springing up along a roadside wall, and he conjectures that the seeds must have been blown from Hubbard's Grove, "some fifty rods east." He also notes a wet and brushy meadow being "rapidly filled with white pines whose seeds must have been blown an equal distance" (*J* XIV:220). Then follows the passage on "persevering Nature" and, at the end of the Journal entry, the concluding sentence, which comes second in the "persevering Nature" portrait: "It does not imply any remarkable rapidity or success in Nature's operations" (XIV:220–21).

The observation and the personifying language with which Thoreau makes it do not automatically evoke Darwin, but in the following paragraph of *The Dispersion of Seeds,* Thoreau moves more clearly toward current scientific language. At first, personification continues to hold sway. Nature acts in a "haphazard manner," but she "surely creates you a forest at last, though as if it were the last thing she were thinking of. By seemingly feeble and stealthy steps—by a geologic pace—she gets over the greatest distances and accomplishes her greatest results" (*Faith* 36). The phrase "geologic pace" introduces the scientific viewpoint, and Thoreau immediately criticizes the "vulgar prejudice that such forests are 'spontaneously generated.'" Rather than such magical thinking, he personifies science, as if it were speaking for Nature: "Science knows that there has not been a sudden new creation in their case but a steady progress according to existing laws, that they came from seeds—that is, are the result of causes still in operation, though we may not be aware that they are operating" (36).

The source of this passage is the Journal entry of 14 January 1861 (*J* XIV:311–12). The two paragraphs in the Journal become four paragraphs in *The Dispersion of Seeds,* but the order of sentences is similar in the two texts (*Faith* 36–37). In both, Thoreau focuses on the slow and sure processes by which nature gives rise to trees and forests, for slowness, as he remarks in the Journal, answers "to the age and strength of trees" (*J* XIV:312). Nor is reproduction the only use of seeds: "If every acorn of this year's crop is destroyed, or the pines bear no seed, never fear. [Nature] has more years to come. It is not necessary that a pine or an oak should bear fruit every year, as it is that a pea vine should" (*Faith* 37; *J* XIV:312). In the Journal, he concludes, "So, botanically, the greatest changes in the landscape are produced more gradually than we expected" (XIV:312). In *The Dispersion of Seeds*, he adds a final paragraph, returning to the idea of small beginnings giving rise to great changes: "However, Nature is not always slow in raising pine woods even to our senses. You have all seen how rapidly, sometimes almost unaccountably, the young white pines spring up in a pasture or clearing. Small forests thus planted soon alter the face of the landscape. Last year perhaps you observed a few little trees there, but next year you find a forest" (*Faith* 37). Here, Thoreau nearly echoes the Journal phrase "botanically, the greatest changes in the landscape are produced more gradually than we expected," but he shifts the idea to the unaccountably *rapid* altering of "the face of the landscape." Whether gradual or rapid, these unaccountable changes are in fact the result of the constant, persistent dispersion of seeds, not the sudden and spontaneous creation of a forest. As the Journal entries of 7 November 1860 and 14 January 1861 demonstrate, Thoreau's thinking and writing echo the "persevering" quality of nature, especially in its botanical forms and processes.

Thoreau returns to the question of spontaneous generation in the discussion of wingless seeds (*Faith* 67–68), focusing on the role of birds in dispersing the seeds of fruit trees and shrubs. In the later pages of chapter 1, Thoreau turns to "downy-seeded herbs" that are dispersed by wind (82–95). The specific case of fireweed raises the question of spontaneous generation directly, for Thoreau had, after publishing "The Succession of Forest Trees," corrected Horace Greeley on the subject in the pages of the *New-York Weekly Tribune*.[6] In *The Dispersion of Seeds*, he mentions a later correspondent of the *Tribune* who notices the dense profusion of downy seeds from fireweed, and that leads Thoreau to note the geographical distribution of plants. If fireweed is spontaneously generated in America, why not in Europe? And if Canada thistle is spontaneously generated in Europe, why then did it not appear in America until the seed had come over from Europe? (88–90).

Debunking a popular misconception, even if it is promoted by his literary agent, is not nearly as important in Thoreau's argument as supporting and applying Darwin's theory of evolution by natural selection. At the end of chapter 1, he takes his support a step further, making the "development theory" his own. As always, Thoreau bases his argument in observations. Digging a pond anywhere in the fields of the Concord area, he notes, you will soon have all manner of water plants, waterfowl, fish, and reptiles in your pond. One example is the Sleepy Hollow cemetery pond, which he had surveyed for the town in August 1855. By 1859, the work of construction was finished, and by 1860 the artificial pond was full of lilies and other aquatic plants (*Faith* 100–101). The Journal of 10 October 1860 is the source for the Sleepy Hollow example (*J* XIV:109–10). Another, closely related example comes directly from the Journal of 18 October 1860, during the period of Thoreau's most intense study of seed dispersion and species succession in surrounding woodlots. He notes that a little pool in Beck Stow's swamp now contains "spatterdock pads and *Pontederia*" both common aquatic plants. But the common plants raise a fundamental question: "How did they get there?" (*Faith* 101). No stream flows there, so he hypothesizes that reptiles and birds must have transported seeds. Since spontaneous generation or "new creation" is not an option, he remarks with some humor, "*Indeed* we might as well ask how they got anywhere, for all the pools and fields have been stocked thus, and we are not to suppose as many new creations as pools" (101).

In a move reminiscent of the earth seed analogy, Thoreau imagines all the pools in the area having been stocked with lilies ages ago, always in the same manner they are stocked now, and that the variations among lily species result from their geographic isolation. His argument draws from Charles Lyell's *Principles of Geology* (1830–33), which Thoreau first read in 1840 and whose uniformitarian principle he recorded in an early page of his Journal: "We discover the causes of all past changes in the present invariable order of the universe."[7] As he puts it in *The Dispersion of Seeds*, "We see thus how the fossil lilies, if the geologist has detected any, were dispersed, as well as those which we carry in our hands to church. Unless you can show me the pool where the lily was created, I shall believe that the oldest of the fossil ones originated in their locality, in a similar manner to these at Beck Stow's" (*Faith* 102).

Darwin's work in the Galápagos and his description of species variations on the islands in *Voyage of the Beagle* play a substantial role here as well. In focusing on the uniformitarian principle of Lyell, Thoreau invokes variations within species, just as Darwin had done in working toward his ideas of island biogeography. As many commentators have noted, Darwin's

finches are the classic example of adaptive radiation, a means of speciation through natural selection. Thoreau echoes the insight in his view of
lily species diverging in various pools or ponds: "Yet I have no doubt that
peculiarities more or less considerable have thus been gradually produced
in the lilies thus planted in various pools, in consequence of their various
conditions, though they all came originally from one seed" (*Faith* 101).
Geographical isolation, archipelago speciation, adaptive radiation, natural
selection—Thoreau is very close to insights made most famous by Alfred
Russel Wallace and Charles Darwin.[8]

If Lyell is the geologist lurking in Thoreau's analysis of pond lilies, Darwin is the botanist behind his vision of the "development theory" (*Faith*
102). Thoreau quotes Darwin's *Origin of Species* in direct support of his explanation of the dispersion of lily seeds by birds. Darwin recounts taking
three tablespoonfuls of mud from a little pond, keeping the mud covered
in his study for six months, and pulling up and counting plants as they ger-
minated—in all, 537 plants, "'and yet the viscid mud was all contained in
a breakfast cup! Considering these facts I think it would be an inexplicable
circumstance if water-birds did not transport the seeds of the fresh-water
plants to vast distances'" (102). The larger implications of Thoreau's reading
of Lyell and Darwin come out in his own statements. "The development
theory," he writes, "implies a greater vital force in Nature, because it is
more flexible and accommodating, and equivalent to a sort of constant new
creation" (*Faith* 102; *J* XIV:147). Or in this botanical application of Lyell's
geological principle: "We find ourselves in a world that is already planted,
but is also still being planted as at first" (*Faith* 101; *J* XIV:146).

A GRANARY AND A SEMINARY

The Journal entries of October–November 1860 play a large role in chapter 2 of *The Dispersion of Seeds* (104–51). As he assembled and wrestled with
the material for the chapter, Thoreau began to figure a way of presenting
his faithful record. The first paragraph suggests the complications of his
method: "I do not always state the facts exactly in the order in which they
were observed, but select out of my numerous observations extended over
a series of years the most important ones, and describe them in a natural
order" (*Faith* 104). The source for this important sentence occurs in the Journal entry for 31 October 1860 (*J* XIV:199). It suggests that at the end of October 1860, Thoreau is casting his eyes back over the accumulated excursions

and the observations made over several years, coming to a sense of the most important "facts" (a favorite term in Darwin's *Origin of Species*) and attempting to find a "natural order" in which to present the material. Perhaps "natural order" is an echo of the botanical natural system of classification, but the order of ideas in the chapter appears to be incompletely realized.

Because "Succession of Forest Trees" contributes a significant number of pertinent passages, we might expect that chapter 2 would be an orderly expansion of the September 1860 address and its published versions. But, in addition to the role of "Succession of Forest Trees," the large number of Journal entries from October 1860 complicates Thoreau's argument and enriches the presentation of his experiences in the woodlots around Concord.

The role of "Succession of Forest Trees" emerges from eight different parts of chapter 2, in which Thoreau transposes sections of the address to *The Dispersion of Seeds*. It is quite clear, to begin with, that the first chapter establishes the widespread methods of transporting and planting seeds by animals and adds the fundamental roles of wind and water to the agencies of quadrupeds and birds. The second chapter is a "natural" or logical extension of the first one; the dispersion of seeds grounds the succession of species. The connections between the address and *Dispersion* are already suggested in the two paragraphs from "Succession" that introduce chapter 1 (*Faith* 24), but they become even more significant in chapter 2.

In general, Thoreau tends to transpose several paragraphs of "Succession" into the appropriate discussion in chapter 2. The first passage, for example, recounts an excursion down the Assabet River on 24 September 1857 and a walk to a dense pine lot that same day. Both parts of the excursion appear in the Journal for that date (*J* X:40–41), and both are transposed directly to *Dispersion* from "Succession" (*Exc* 171–72). This excursion is significant because it registers Thoreau's surprise at finding "my theory so perfectly proved in this case," the theory being that squirrels are constantly planting the nuts of hickory and oak trees (*Faith* 108).

This method of transposing whole sequences of paragraphs from "Succession" to *Dispersion* persists in all eight parts of the later work. The passages from "Succession" do not appear in the same sequence in *Dispersion*, but the order of ideas in the two texts is quite similar. The second passage, for instance, jumps to later paragraphs of "Succession," bolstering Thoreau's discussion of the vitality of acorns and other seeds that lie dormant in the soil (*Exc* 179–81; *Faith* 111–14). The six paragraphs of this discussion feature Thoreau's critique of previous botanists who had promoted the idea of "the tenacity of life" in various seeds (*Faith* 112). Thoreau accepts the idea provisionally, citing excursions to the demolition and ruins of the Hunt House in 1859 and 1860 as examples for discovering "long-extinct plants" springing from the site (113). He concludes, nonetheless, that despite the exceptions the primary means of dispersion is by animals (114).

The third and fourth passages follow closely on the discussion of vitality of seeds, both in "Succession" and in *Dispersion*. Both appear in one paragraph of "Succession," focusing on the role of jays in dispersing acorns in the autumn and citing William Bartram as an authority (*Exc* 178–79). In *Dispersion,* Thoreau separates the experience of hearing jays pecking at acorns from the quotation of Bartram (*Faith* 114–15, 116). One reason for the separation is that excursions of 27 and 29 October 1860 intervene, with more evidence of the role of jays in disseminating acorns (115–16). Bartram functions to conclude the discussion of jays and confirm the accuracy of Thoreau's own observations and inferences. This play of sources creates a kind of polyphony of records and voices, in which Thoreau's excursions and his faithful recording of them engage directly with the authority of William Bartram. Indeed, Thoreau repeatedly includes the voices of jays and squirrels as well, giving their voices considerable play in the dialogues he creates.

After these transpositions, we find four more passages in a nearly linear sequence of paragraphs from "Succession" (*Exc* 173–74; 174–75; 175–77; 177–78, 174). Much intervening material from October 1860 excursions breaks up the sequence of paragraphs, and the effect is that Thoreau disperses the "Succession" passages in strategic locations of *Dispersion*. Most of these passages, in fact, focus on the role of animals—especially squirrels—in transporting and planting acorns, chestnuts, and hickory nuts ("pignuts," as Thoreau calls them). They show the persistent and widespread role of animals in disseminating the seeds of trees, ending with a sense of the immense quantity of seeds that squirrels collect and plant (*Faith* 142–43). In sum, the eight passages from "Succession" provide a kind of narrative spine and argumentative structure for chapter 2. The sequence of topics is arranged roughly by species, with oaks, chestnuts, and hickories being the principal discussions.

After the last transposition from "Succession," in fact, chapter 2 of *Dispersion* tends to lose focus and coherence. The final eight pages (*Faith* 143–51) are brief, scattered notes, rather than developed commentaries or narratives of excursions, and they range from shrub oaks and hazelnuts to the role of mice and common resident birds in transporting seeds, especially acorns.

The "natural order" that Thoreau announces in the opening of chapter 2 does not come fully to fruition. As in chapter 1, he focuses most sharply on the relationship of acorns and nuts to the squirrels and birds that disperse them, but the relationship to chapter 1 and to "Succession of Forest Trees" is by no means the whole story of this chapter.

The Journal materials from October 1860 take meaningful steps toward a natural order for chapter 2. They show how well Thoreau incorporates the excursions into his treatment of succession, and how much he discovers about the processes of succession after presenting "Succession of Forest Trees." The excursions of October 1860 allow Thoreau to advance his analysis of succession by stating a research problem: "At first thought, one would expect to find seedling oaks in the greatest abundance, if not exclusively, under and about seed-bearing oaks, that is, in oak woods; but when I looked for them there, they were obviously fewer and feebler than under pines" (*Faith* 109). To account for this key observation, Thoreau makes further excursions to compare oak seedlings, and this study of oak seedlings forms the core of the chapter (109–25).

Once that study is delineated, Thoreau turns to related studies of chestnut seedlings (*Faith* 126–31) and hickory seedlings (136–43). In the section on hickories, Thoreau uses a Journal entry from 24 September 1860 (*J* XIV:93–94) to praise "how much vitality there is in the stumps and roots of some trees, though small and young," contrasting that vitality to the repeated cutting down of seedlings to keep a pasture clear (*Faith* 136–37). In the Journal, the description of the hickory seedlings leads Thoreau to bemoan "how heedlessly woodlots are managed at present" (*J* XIV:94), but he holds most of his critique of forest management for chapter 3 of *Dispersion*. A rather prolonged discussion praising squirrels (*Faith* 131–36) breaks up the sequence of three seedling studies, but Thoreau's clear-eyed affection for the animals keeps the digression from being a distraction. This is also a passage in which he interpolates several pages from "Succession" (*Exc* 175–77).

The oak seedling study begins on 27 October 1860. Thoreau makes an excursion to collect ten seedling oaks in purely oak woods and another ten seedling oaks in purely pine woods (*J* XIV:179; *Faith* 109). In the Journal, Thoreau details his search, recounting the time spent searching in at least four woodlots (*J* XIV:178–83). The collection is less than wholly successful, for he finds few oak seedlings in oak woods, while there are countless oak seedlings under the pines (*Faith* 109–10). Thoreau draws a preliminary conclusion, in *Dispersion*, that pine woods may be "a natural nursery of oaks" (110). That preliminary conclusion stems from the 17 October 1860 Journal (*J* XIV:139). In the 27 October Journal entry, he goes much further in his conclusions, but these appear later in *Dispersion*. After an interval of "Succession" material (*Faith* 111–15), Thoreau returns to the 27 October excursion to note the presence of jays as planters of acorns, adding a passage from the Journal entry for 29 October as further evidence (*J* XIV:180, 188; *Faith* 115). The jays seem to intervene here partly because Thoreau encounters a scolding jay during the excursions of 27 and 29 October, not because they are part of a "natural order" of ideas. As in other passages about jays and squirrels, the animals and their voices seem to pull at the writer's attention, and they play a significant role in the faithful record of his excursions.

This kind of easy digression is in fact a pattern in Thoreau's prose. In the present case, for example, he is not keenly focused on the ten seedlings of his initial study. He recounts excursions of 17 October 1860, 16 October 1860, and 24 October 1860, all of which yield interesting observations about oak seedlings, the lengths and shapes of their roots and shoots, and the relative health of seedlings found under oak woods and under pines (*J* XIV:139, 128–29, 169–70). While the presentation in *The Dispersion of Seeds* clearly moves toward the correlation between the location of oak seedlings and their relative vitality, the Journal entries are by no means presented in chronological order. Indeed, Thoreau moves among the three excursions several times, reversing the order of events and passages in developing his observations about the shoots and roots of oak seedlings (*Faith* 116–19). In the prevailing prose pattern that emerges—one that we have seen repeatedly—Thoreau uses successive layers of material, not always in chronological order, to develop his ideas. Further, the excursion narratives give him opportunities to dramatize his discoveries, even if they jostle the logical sequence of his arguments.

The three excursions of 16, 17, and 24 October serve as detailed background for the study of ten oak seedlings, begun on 27 October 1860. Thoreau returns to this initial topic to develop the comparison of oak seedlings from oak and pine woodlots. In *The Dispersion of Seeds*, he presents the comparison as a methodical process of measurement and description, leading to three conclusions as to why oak seedlings growing in an oak wood are so few and so diseased (*Faith* 119–21). Interestingly, in the Journal, Thoreau leaps to the causal conclusions (*J* XIV:181) *before* he details his careful comparisons (XIV:181–83). In the Journal, the conclusions in fact function as hypotheses that he tests with the comparative analysis. In *The Dispersion of Seeds*, the inductive process becomes a deductive argument.

The "natural order" of chapter 2 allows Thoreau to move fluidly from one excursion to another, merging materials in ways that support his ideas about the succession of oaks and pines and the vitality of oak seedlings in propagating forests. Even though we might expect the comparison of oak seedlings to dominate the discussion, for instance, the excursion of 17 October 1860 figures largely in the presentation. The 17 October 1860 Journal entry runs some nine pages and recounts an afternoon excursion to Walden woods. During the excursion, Thoreau makes a variety of observations about woodlots in the area, seedlings he collects, and the vitality of seedlings in pine woods. The 17 October excursion becomes the focus, then, because of the writer's interest in answering another research question: How long do oak seedlings live in dense pine woods? (*Faith* 121–23). To answer this question, Thoreau collects oak seedlings in several pine woodlots, both pitch and white, using annual rings to compare measurements and ages; he concludes that young oaks survive about ten years, while the squirrels and jays are planting acorns every year (*Faith* 125). Along the way, he interpolates an excursion to John Hosmer's woods on 30 October 1860 and the "Succession" material concerning English methods of using pines as nurse plants for oaks (*Exc* 173–74; *Faith* 122–25).

Without any doubt, Thoreau's successive, cumulative method of writing can be challenging for a reader. While it is clear that he is extremely careful in composing his essays, the transposition of narrative Journal entries into discursive essay forms does not always create strong patterns of coherence. The narrative elements of dates, times of day, locations, actions, observations, and reflections enliven the exposition of botanical and ecological discoveries. On the other hand, it is not always apparent what day or afternoon or month or even year we are reading about. The woodlots are numerous and less than familiar to us, even if we have spent months or years reading about them in the Journal and elsewhere, and even if we have studied the maps of the area provided in scholarly publications of Thoreau's works. The study of oak seedlings does not maintain a dominant role in the exposition; instead, it is presented, then recedes for several pages, then reemerges as the chapter proceeds. The effect is to produce uncertainty in the reader, but I think this uncertainty is productive. The movements from one excursion to another, without following a chronological order, make both the excursions and Thoreau's prose into sites of constant discovery.

Thoreau's "natural order" is thematic as much as botanical or ecological. Chapters 1 and 2 gather species together in significant groups, detailing the processes by which species disperse seeds and create successive populations of trees in a forest. Overall, the two chapters form a logical, argumentative sequence, too, since the processes of dispersion of seeds lead us to questions (and answers) about population succession. More important, they lead, at the end of chapter 2, to the following large conclusion: "The consequence of all this activity of the animals and of the elements in transporting seeds is that almost every part of the earth's surface is filled with seeds or vivacious roots of seedlings of various kinds, and in some cases probably seeds are dug up from far below the surface which still retain their vitality. The very earth itself is a granary and a seminary, so that to some minds its surface is regarded as the cuticle of one great living creature" (*Faith* 151). Both "granary" and "seminary" retain their primary senses of a storehouse of grain or seed-plot (*OED*); both words function figuratively at the same time, converting the earth into a repository of essential vitality. This is once again the earth as *phusis*, an embodiment of a fundamental generative power. At first, we might see the "one great living creature" as a personified animal, but "cuticle" suggests a botanical meaning, the thin epidermal layer covering a seed. Among other possibilities, the passage appears to echo chapter 1 of *The Dispersion of Seeds*, the image of the earth as springing from "a seed as small in proportion as the seed of a willow is compared with a large willow tree" (*Faith* 67). As we have seen, this is an image of "small beginnings" (66) giving rise to mighty trees—or even to the earth itself. The passage may recur just as tellingly to the sand foliage passage in *Walden:* "The earth is not a mere fragment of dead history, stratum upon stratum like the leaves of a book, to be studied by geologists and antiquaries chiefly, but living poetry like the leaves of a tree, which precede flowers and fruit—not a fossil earth, but a living earth; compared with whose great central life all animal and vegetable life is merely parasitic" (*W* 309). This set of figurative relationships shows that the smallest seed can deliver the largest and most faithful discoveries.

UNROLLING THE ROTTEN PAPYRUS

The shortest and last of the three chapters in *The Dispersion of Seeds*, chapter 3 (151–73), engages the human dimension of forests most profoundly. Thoreau uses his excursions to ground an analytical history of woodlots in the Concord area, and this history of land use grounds his critique of the forest management practiced by his neighbors. Much of chapter 3 is taken from the October–November 1860 Journal; as in the first two chapters, the excursions appear in nonchronological sequence, forming a temporal mosaic that corresponds roughly to the mosaic of woodlots and Thoreau's mosaic of concerns. For example, the chapter begins with excursions to seventeen pitch-pine groves, and in thirteen Thoreau discovers the "rule" that "under a dense pitch-pine wood, you will find few if any little pitch pines, but even though there may be no seed-bearing white pines in the wood, plenty of little white pines" (*Faith* 153). The evidence for this rule comes from excursions made on 29 and 30 October 1860 (*J* XIV:187–98), although the two days do not include visits to all seventeen dense pitch-pine woodlots.

The focus on pitch pines takes up the first section of the chapter (*Faith* 151–58), and along the way Thoreau delineates some of the principal features of pitch-pine woodlots. They tend to grow on bare ground or old pastures, and often on poor, sandy soil. They are more light dependent than white pines, and therefore they tend to take root on the edges of existing woodlots of other species. In open pastures, they encourage other tree species, such as white pines, birches, shrub oak, other oaks, and even apple trees to find a place; thus, they support the development of mixed woods. As for their original growth patterns, before white settlers colonized the area, Thoreau conjectures that Native Americans used fires and clearings to make bare plains conducive to pitch pine groves.

White pines, Thoreau discovers, tend to be more widely dispersed than pitch pines and form mixed woods more often than the pitch pine (*Faith* 155). When he turns his discussion to white pine woodlots, he notes that he has found only three dense white-pine groves, and that they do not favor the growth of white pine seedlings (158–59). Thoreau finds more little white pines springing up in oak woods. Investigating this phenomenon, he makes repeated excursions to the three oldest oak woods in the area (Wetherbee's, Blood's, and Inches), visits that occupy him for much of November 1860. The first observation is that white pines are distributed widely in all three old oak woodlots. Thoreau draws the interesting inference that natural succession is taking place, and that the succession from oak to white pine is a natural pattern for "primitive" woods (160–61). On the heels of this discussion, Thoreau states a larger pattern of succession:

> We have seen how the white pine succeeds to the pitch pine quite commonly, just as oaks succeed to pines generally. So it will succeed to white pine (or itself) when a white-pine wood which is more or less open, and

hence contains plenty of little pines, is cut, though there will probably be oaks mixed with them in that case unless the little pines are thick and considerably grown. So it may succeed to an oak wood in which it has already sprung up thickly, especially to an old oak wood whose stumps fail to send up sprouts. It may spring up as soon as such an oak wood is cut, the ground being bare, but this depends on circumstances. (161)

The "circumstances" may be the number of seeds borne by white pines or white oaks in a given area. The excursion of 16 October 1860 provides an example of a mixed woodlot of large white pines and oaks, cut off the previous winter. Thoreau discovers twenty white-pine seedlings immediately, but not a single oak seedling of that year (161). In the Journal, the record of this observation appears in a brief paragraph, and *The Dispersion of Seeds* formulates the larger pattern of succession by elaborating on the brief inference: "This shows how much the species of the succeeding forest may depend on whether the trees were fertile the year before they were cut, or not" (*J* XIV:131–32).

The excursion of 16 October 1860 may be one of the most fruitful that Thoreau ever made. The entry appears and reappears in the pages of chapter 3, yielding insight upon insight. The excursion, building on hundreds of earlier excursions and dozens of subsequent ones, clearly gives Thoreau a deep knowledge of land-use history. Both the Journal and *The Dispersion of Seeds* deliver that accumulating knowledge in a faithful record: "Our woodlots, of course, have a history, and we may often recover it for a hundred years back, though we *do* not. A small pine lot may be a side of such an oval as I have described, or a half or a square in the inside with all the curving sides sheared off by fences. Yet if we attended more to the history of our woodlots, we should manage them more wisely" (*Faith* 164). The observation about pines growing in "regular rounded or oval or conical patches" comes at the beginning of the excursion (*J* XIV:125), and this growth pattern is due to the dispersion of winged pine seeds by wind (XIV:124–25). In *The Dispersion of Seeds,* Thoreau contrasts the "wild woods" and the "regular progress of things" with "settled countries like this," in which owners cut and plow up "successive crops of trees or seeds" before allowing "Nature to have her way" (*Faith* 164). That contrast comes only at the end of the 16 October Journal entry (*J* XIV:133), but by placing it just before the paragraph about woodlots having a history, Thoreau emphasizes the ways in which owners disrupt the "regular progress of things." By reading the natural history in the growth patterns of woodlots, moreover, we may learn how we "should manage our forests more wisely."

The history of woodlots is precisely the topic that Thoreau develops most fully in the 16 October 1860 Journal entry. The record of the excursion moves from concrete descriptions of woodlots to large-scale inferences drawn from the "forest geometry" and the problems it presents (*J* XIV:127). In *The Dispersion of Seeds,* Thoreau transposes some eleven paragraphs of the Journal in near-linear sequence, as if he were solving his own geometrical writing problems by allowing the material to take its shape from the "regular progress of things." And he succeeds, for this section on woodlot histories effectively describes the ways in which Thoreau reads the patterns and discovers the history underlying them.

The style of writing in *The Dispersion of Seeds* dramatizes discovery. For example, Thoreau opens the eleven-paragraph section by setting the scene: "This afternoon, in the middle of October, as I am walking across the fields in the outskirts of the town, I observe at a distance an oak woodlot some twenty years old" (*Faith* 165). The Journal begins the paragraph simply: "Looking round, I observe at a distance" (*J* XIV:126). The careful use of deictics, placing the narrator in time and space, emphasizes the role of the walker in investigating the history of the woodlots around him. The long paragraph in the Journal becomes three paragraphs in *The Dispersion of Seeds,* clarifying a significant progression of ideas. Reading the history of the oak woodlot, the walker discovers a deeper history of a previous pitch-pine wood, whose seeds were dispersed before the pines were cut (*Faith* 165). That explains a "dense, narrow edging of pitch pines" along the margin of the oak woodlot, and it leads to the new sentence that Thoreau writes to lead the third paragraph: "This is the history of countless woodlots hereabouts, and broad ones, too. But, I ask, if the neighbor so often lets this narrow edging grow up, why not oftener, by the same rule, let them spread over the whole of his field? When at length he sees how they have grown, does he not often regret that he did not do so? Or why be dependent, even to this extent, on these windfalls from our neighbor's trees, or on accident? Why not control our own woods and destiny more?" (165–66). The history delineated in the first two paragraphs leads to the questions of forest management in the third. As he digs deeper into the history and draws more profound inferences concerning the practices of the owners of the woodlots, Thoreau develops a narrative that explains the "forest geometry" before his eyes.

"But," the writer of *The Dispersion of Seeds* tells us, "I have not done with the last-named woodlot yet" (*Faith* 168). He uses passages from the 19 October 1860 excursion to flesh out the image of this "double forest" of pines and oaks (167), and he recounts a further discovery, that an "old oak wood" had stood on the property before either the pitch pines or the current oaks had been planted (*J* XIV:150–51). This deeper history leads Thoreau to discern "three successions of trees, and, I may say, five generations," for the history stretches forward to the "strip of little pines" and "oak seedlings under the little pines" (*Faith* 169). "Thus," the writer concludes, "you can unroll the rotten papyrus on which the history of the Concord forest is written" (*J* XIV:152; *Faith* 169).

The papyrus is rotten in at least two ways. First, the stumps of a large pine wood, cut some fifteen years before, give Thoreau part of the history he tells, and the stumps of an even older oak wood, cut between fifty and sixty years earlier, take him farther into the past. Some of the stumps he discovers in his excursions could lead even farther back to "before the white man came" (*Faith* 169). Therefore, the stumps and their growth rings, decaying as many of them are, are the rotten papyrus we can read, taking us from the present to ancient times. But the papyrus may be rotten, too, because it suggests all the unthinking, foolish, and misguided practices that produced the stumps.

In the last four pages of *The Dispersion of Seeds*, Thoreau draws on the Journal entries for 16, 17, and 19 October 1860 to deliver a stinging rebuke to the landowners of the Concord forests. The 16 October excursion is especially fruitful in that regard. Thoreau characterizes "the history of a woodlot" as "often, if not commonly, *here* a history of cross-purposes." In the Journal and in *The Dispersion of Seeds*, the conflict is between the "steady and consistent endeavor on the part of Nature," an image of uniform perseverance that Thoreau repeats in all three chapters of the book, and the "interference and blundering with a glimmering of intelligence at the eleventh hour on the part of the proprietor" (*J* XIV:132; *Faith* 170). On the excursions, and faithfully recorded in his writing, Thoreau is outraged by the ignorant, reckless way in which a proprietor can treat the "gift of a forest" that springs up in a "starved pasture" without his notice (*J* XIV:133; *Faith* 171). The battle between woodlots and pastures appears in the 17 October entry, and the tendency of landowners to let their cattle roam in woodlots appears in the 19 October entry (*J* XIV:145–46, 150–51). The result of this "actual history of a great many of our woodlots" is that we have "both poor pastures and poor forests" (*Faith* 172).

In the final six paragraphs of *The Dispersion of Seeds* (*Faith* 172–73), Thoreau fashions a harsh peroration out of two paragraphs from the 16 October Journal entry (*J* XIV:130–31). He narrates the excursion to a distant part of the town to visit a dense white pine wood, cut the previous winter. He intends merely to "see how the little oaks, with which I knew the ground must be filled there, looked now," but he finds "that the fellow who calls himself its owner has burned it all over and sowed winter rye there!" Then, another outburst: "What a fool!" The text in *The Dispersion of Seeds* shows a great deal of care in Thoreau's revisions of the Journal account. He rearranges sentences, reorders the sequence of paragraphs, and adds an image to emphasize the steady designs of personified Nature: "A greediness that defeats its own ends, for Nature cannot now pursue the way she had entered upon" (*Faith* 173). Without belaboring the point by prosecuting a micro-analysis of the style, we can observe how Thoreau develops the withering contrast between the bumbling, greedy owner, with his view of the land as "pine-sick," and an always-prepared, ever-ready, vital force embodied in the woodlot that has been cut, burned, and sown with winter rye. As he closes his draft of the incomplete book, Thoreau completes his scathing argument for the well-being of the people and forests of Concord and indignantly condemns the owners to a servile state: "Forest wardens should be appointed by the town—overseers of poor husbandmen" (173).

CHAPTER 6

Wild Fruits and Transformative Perceptions

In 2000, Bradley Dean published a near-miraculous book, *Wild Fruits,* edited from the manuscript "Notes on Fruits," some 306 leaves of material in the Berg Collection at the New York Public Library. Two of Dean's miracles were finding a reasonable phenological order of entries and deciphering the layered, incremental revisions in Thoreau's notoriously difficult handwriting.[1] Not surprisingly, Dean interpreted *Wild Fruits* as part of the larger Kalendar project, which could have become a "comprehensive history of the natural phenomena that took place in his hometown each year" (WF xi). The late essay "Wild Apples" is a major part of the *Wild Fruits* project, and the consensus among scholars is that Thoreau used the material as a way of publishing at least part of the larger work before his death.[2] If "Wild Apples" is related to *Wild Fruits* in a synecdochic way, it could resemble the relationship between the "Succession of Forest Trees" lecture and the *Dispersion of Seeds* project. Despite the possible parallel, the "Wild Apples" essay seems more of an important part among many parts than a microcosm of the entire *Wild Fruits* project. Other entries such as "Black huckleberry" and "European cranberry" add nuance to the project and its interpretation.

Though by no means complete, *Wild Fruits* is a dynamic, performative hybrid of many fragments, an experimental form of writing based on repeated excursions and observations of plants. It gives the most developed example of how Thoreau's botanical work of the last decade delivers new thinking and writing. Thoreau's botany proposes a more intimate relationship between human beings and the land they occupy, manage, and cultivate. *Wild Fruits* is persistently unsettling, merging wild and cultivated fruits to develop fresh ideas of growth and transformation. The theme of intimate relationship connects *Wild Fruits* to the *Dispersion of Seeds* manuscript, especially the final chapter on forest management.

The order of the entries, established by Dean in editing the manuscript, is phenological. The 181 entries are given in the order of first fruiting, beginning with the elm tree, which fruits between 7 and 9 May, and ending with *Juniper repens,* which fruits around 1 March. This makes *Wild Fruits* an actual book-length avatar of the Kalendar project. In addition, the seasonal organization resonates significantly with the structure of *Walden.* The temptation may be to read the seasonal text as a stable, timeless artifact; as in *Walden,* however, Thoreau embeds dynamic elements in *Wild Fruits,* using himself and his excursions as a way of encountering plants and rendering the encounters as vital, potentially transformative experiences. In *Wild Fruits,* we once again read the botanical excursion as a way of knowing and the repeated image of the walker as a figure of knowing, thinking, and experiencing. Moreover, Thoreau persists in understanding his writing as experiential, including the experience of constant regeneration.

Wild Fruits is clearly a work in progress, but it displays the structural characteristics of a skeletal draft, an adumbration of the unfinished final work, rather than a collection of notes. Fifteen entries are longer than five pages, and these tend to focus on some of the most common and important fruits in New England. In addition, a half dozen entries are two to four pages long, and in several cases the draft entry is likely designed to become a short essay. In some cases, a short entry may function as a preliminary sketch for a longer essay. A large, third category is comprised of entries, of varying length, that simply compile observations from the Journal. How these entries might have developed into short or long essays is impossible to know. In all three categories, the Journal functions as Thoreau's principal source for the manuscript and the materials compiled there.

The fifteen entries in the first category include two versions of an introduction (WF 3–6, 242–44) and an untitled conclusion (233–39). The remaining "long" essays, in order of fruiting, are "Strawberry" (10–17), "Early low blueberry" (21–26), "High blueberry" (30–36), "Black huckleberry" (37–59), "Wild apples" (74–92), "Common cranberry" (102–7), "Barberry" (139–43), "Wild grape" (150–57), "European cranberry" (164–70), "Acorns" (178–86), "Chestnut" (209–16), "Walnuts" (216–20), and "Pitch pine" (227–32). By length, the main entries are "Strawberry" (8 pages); "Black huckleberry" (23 pages); "Wild apples" (19 pages); "Wild grape" (8 pages); "Acorns" (9 pages); and "Chestnut" (8 pages). The others are 5–7 pages long.

The plants observed range widely, from large trees to ground nuts. "Fruits" is an inclusive botanical term; in Thoreau's botany, it functions most importantly as a value-laden figure. According to Thoreau, all plants produce valuable fruits. In *Wild Fruits*, the dandelion is chronicled as well as the blueberry, though Thoreau is drawn to wild fruits that are edible and generally considered delicious. Even the term "wild" is broadly understood. Substantial entries on domesticated fruits include "Watermelons" (*WF* 107–10) and "Potatoes" (115–19); shorter entries appear for "Grains," "Turnip," "Muskmelons," "Pear," "Peach," "Pumpkins," "Peas," "Beans," and "Corn."

Taken together, the two alternate introductions to *Wild Fruits* present a clear set of concerns for the book. The longer of the two, some ten paragraphs, is drawn from the Journal of 22 November 1860 (*J* XIV:261–62), 26 November 1860 (XIV:273–74), and 28 November 1860 (XIV:277–78). Both in chronology and in thematic development, the material comprising the introductions is closely related to the *Dispersion of Seeds* project. As the editor of *Wild Fruits* notes, the Journal and the two book projects display a host of complex, intricate connections.[3]

Both introductions focus on establishing the interest of wild fruits as a topic for a book-length study. In material drawn from the Journal of 7 March 1859 (*J* XII:23–24), Thoreau proposes a fundamentally metaphorical understanding of the term "fruit." He claims to use it "in the popular sense," but he immediately insists on a deep, layered meaning for the word: "The ultimate expression or *fruit* of any created thing is a fine effluence, which only the most ingenuous worshipper perceives at a reverent distance from its surface even. The cause and the effect are equally evanescent and intangible, and the former must be investigated in the same spirit and with the same reverence with which the latter is perceived. Only that intellect makes any progress toward conceiving of the essence which at the same time perceives the effluence" (*WF* 242). The word "effluence" signifies an outpouring or outflowing, a dynamic image for the fruit or expression of a plant—or any other created thing. In this introduction, Thoreau focuses on a nonscientific account of wild fruits: "Science is often like the grub, which, though it has nestled in the very germ of the fruit, and so perhaps blighted or consumed it, has never truly tasted it." Indeed, the "mystery of the life of plants is kindred with that of our own lives," and it would be a grave mistake to "probe with our fingers the sanctuary of any life, whether animal or vegetable" (242). The definition of "fruit" stresses the spiritual language of reverence, worship, evanescence, intangibility, and mystery. Likewise, the epistemology of Thoreau's botany emphasizes metaphorical, more-than-human ways of knowing, while it is also fundamentally grounded in the life of plants.

In the Journal entry and in the shorter of the two introductions, a kindred term for "fruit" is "ripeness," a word for which Thoreau insists on a spiritual meaning: "There is no ripeness which is not, so to speak, something ultimate in itself, and not merely a perfected means to what we believe to be a higher end" (WF 243). Thus "the fruit of a tree is neither in the seed, nor the timber—nor is it the full-grown tree itself—but I would prefer to consider it for the present as simply the highest use to which it can be put" (243). The phrase "highest use" immediately recalls Thoreau's praise of the pine in the "Chesuncook" chapter of The Maine Woods. In that famous passage, he asserts that "the pine is no more lumber than man is, and to be made into boards and houses is no more its true and highest use than the truest use of a man is to be cut down and made into manure. There is a higher law affecting our relation to pines as well as to men" (MW 121). The essence and effluence of the pine form "the living spirit of the tree, not its spirit of turpentine" (122).

Thoreau's stand against scientific and industrial materialism continually argues for an alternative theory of value, one that sees commercial, monetary uses and values as plainly inferior to spiritual perceptions and metaphorical meanings. For that argument, he consistently undercuts the "popular sense" of fruits and their value. In the longer of the introductions, for instance, he praises "our native fields" and the discovery of "a new fruit" there, in contrast to the popular value given to cultivating "imported shrubs in our front yards" and the sale of "tropical fruits" or "table fruits" in the marketplace (WF 3). The "new fruit" is local and wild, and its value far outstrips exotic and tropical imports like pineapples and oranges: "The value of these wild fruits is not in the mere possession or eating of them, but in the sight and enjoyment of them. The very derivation of the word 'fruit' would suggest this. It is from the Latin fructus, meaning 'that which is used or enjoyed'"(4). For Thoreau, the process of gathering wild fruits is essential to the value of the fruits. If a child makes an excursion "a-huckleberrying," the excursion introduces the child "into a new world," in which she "experiences a new development." With an assumed air of nonchalance, Thoreau asserts that "the value of any experience is measured, of course, not by the amount of money, but the amount of development we get out of it" (4).

Commerce may exploit and market the objects we commonly call fruits, but for Thoreau commerce seizes "always the very coarsest part of a fruit— the mere bark and rind, in fact, for her hands are very clumsy" (WF 5). The fairest fruits or parts of fruits are not sold and cannot be bought: "you cannot buy the highest use and enjoyment of them. You cannot buy that pleasure which it yields to him who truly plucks it. You cannot buy a good appetite, even. In short, you may buy a servant or slave, but you cannot buy a friend" (5). The pleasures of local fruits lie partly in the gathering of them, in the kinship or friendship they offer human beings, and in the teachings they give us: "Do not think, then, that the fruits of New England are mean and insignificant while those of some foreign land are noble and memorable. Our own, whatever they may be, are far more important to us than any others can be. They educate us and fit us to live here. Better for us is the wild strawberry than the pine-apple, the wild apple than the orange, the chestnut and pignut than the cocoa-nut and almond, and not on account of their flavor merely, but the part they play in our education" (5). Like "fruit" and "ripeness," the word "education" assumes a multivalent metaphorical meaning in this passage: it signifies a process through which inhabitants become "fit" to live in their local place. How does one achieve such "fitness" to a place? As a start, by the repeated excursions in the Concord neighborhood, the disciplined study of plants and their habitats, and the persistent meditation on the relationship between plants and human beings as kindred spirits or friends—the education of Thoreau's readers would entail all these projects of fitness, with the goal of a kind of physical soundness and spiritual health. It would create new ways of knowing, but it would even more importantly create new ways of being in the world.

Another vital aspect of education is the value of experience, measured not by money but by "the amount of development we get out of it" (*WF* 4). The word "development" directly evokes Darwin's *Origin of Species,* which powerfully undergirds Thoreau's thinking in *The Dispersion of Seeds.* Here, too, the "development theory" evokes a fine effluence, as in the 18 October 1860 Journal, a passage brought into the text of *The Dispersion of Seeds* for its great implications: "The development theory implies a greater vital force in nature, because it is more flexible and accommodating, and equivalent to a sort of constant *new* creation" (*J* XIV:147; *Faith* 102). As we have seen in multiple readings from the Journal and other works, Thoreau's excursions amount to "a sort of constant *new* creation." The new appears as a surprising botanical discovery, or an insight into the human relationship to botanical neighbors, or a meditation on the inherent vitality and spirituality of nature itself. The new is in effect "the amount of development" we get out of experience.

Read alongside the Journal entries of March 1859 and October–November 1860, and placed next to the *Dispersion of Seeds* manuscript, the two introductions to *Wild Fruits* express ambitious goals for Thoreau's botany and the writings that develop from it. The text of *Wild Fruits* falls short of meeting those goals, but several entries indicate the fruitful directions Thoreau's writing was taking in his final months.

WILD APPLES AND THE WALKER'S COMFORT

The published essay "Wild Apples" appears in Joseph Moldenhauer's scholarly edition of *Excursions,* which I have cited for "Autumnal Tints," "Walking," and "An Address on the Succession of Forest Trees." For the present discussion, I prefer to cite the text in *Wild Fruits,* partly to keep our attention on that late project, and partly because the *Wild Fruits* version seems to represent Thoreau's writing in nearly original form. Of course, as is often the case in reading Thoreau, there are several texts to choose from.

Thoreau delivered the "Wild Apples" lecture twice. The first time was at the Brick School House in Concord, 8 February 1860; the second was 14 February, six days later, in Bedford (*Exc* 633–34). The Journal was, as usual, his main source, using material from as late as January 1860; he even added material on the crab apple to the essay after his Minnesota trip of May–June 1861 (634). It was the last lecture Thoreau prepared for publication in the *Atlantic Monthly,* and the *Atlantic Monthly* text is the copy text for all printings. Bradley Dean's version clips the introduction and conclusion and omits the section headings, making the essay conform with the other entries in *Wild Fruits* by beginning with the phenological statement "Early apples begin to be ripe about the first of August; but I think that none of them are so good to eat as some to smell" (*WF* 74). Dean prints the introduction and conclusion of the published essay in the section on "Related Passages" (244–48).

Dean's phenological beginning for "Wild Apples" keenly echoes the introduction to *Wild Fruits,* for the entry immediately notes that smell is as good as taste for early apples. As in the introduction, Thoreau emphasizes a sense of value in earthly fruits that transcends the merely material: "There is thus about all natural products a certain volatile and ethereal quality which represents their highest value, and which cannot be vulgarized, or bought and sold" (*WF* 74). He imagines a "particularly mean man carrying a load of fair and fragrant early apples to market," and he observes that "though he gets out from time to time and feels of them and thinks they are all there, I see the stream of their evanescent and celestial qualities going to heaven from his cart, while the pulp and skin and core only are going to market" (75). The strong echoes between this phenological opening and the spiritually focused draft introductions suggest the aptness of Dean's decision to bring "Wild Apples" into line with the other entries of *Wild Fruits*. In addition, the entry is itself organized phenologically, moving from the early apples to the "last gleaning" in mid-November and the "frozen-thawed" apple of late November (*Exc* 285, 286).

Shifting his focus from cultivated to wild apples, Thoreau considers at length "how the wild apple grows," as the *Atlantic Monthly* subheading terms it (*Exc* 272). Thoreau admires the crab apple as America's native and aboriginal fruit, but he associates himself with wild plants that "have strayed into the woods from the cultivated stock" (*WF* 79). These "backwoodsmen among the apple trees" spread through the dispersion of seeds by birds and cows, and they become wild by growing as scrubs and shrubs, despite being repeatedly browsed and "cut down annually" by grazing cattle (80–81). With affectionate humor, Thoreau connects the hardy qualities of wild apples to the idea of education: "Every wild-apple shrub excites our expectation thus, somewhat as every wild child. It is, perhaps, a prince in disguise. What a lesson to man! So are human beings, referred to the highest standard, the celestial fruit which they suggest and aspire to bear, browsed by fate; and only the most persistent and strongest genius defends itself and prevails, sends a tender scion upward at last, and drops its perfect fruit on the ungrateful earth. Poets and philosophers and statesmen thus spring up in the country pastures and outlast the hosts of unoriginal men" (83). This early section of the lecture/essay supports, once again, the reading of "Wild Apples" within the context of *Wild Fruits*. The question of effluence and value is clearly the focus. Just as the cartload of early apples emits a "stream of their evanescent and celestial qualities" that only the few can perceive, so human beings suggest a "celestial fruit" that most of us never bear. At once weary and heroic, Thoreau remarks that "such is always the pursuit of knowledge" (83).

Among the figures for the pursuit of knowledge, the most persistent in the "Wild Apples" entry is the walker, a figure who moves steadily from the earliest pages of the Journal to the last pages of Thoreau's works. In the introductory section of the published essay, Thoreau praises the cultivated apple trees as "delicious to both sight and scent," so much so that "the walker is frequently tempted to turn and linger near some more than usually handsome one, whose blossoms are two-thirds expanded" (*Exc* 264).

Wild apples call for sharper senses, as the walker sometimes lets low apple shrubs go unnoticed (*WF* 81). The farmer imagines that his cultivated apples taste better than the wild ones, but Thoreau is sure that he is mistaken, for he cannot have "a walker's appetite and imagination" (84). The wild apples left out until the first of November belong to "us walkers" (84), Thoreau comments, placing himself rightly among the apple-picking walkers, as opposed to apple-selling farmers.

The wild apple effectively places the walkers within the proper season. "The time for wild apples is the last of October and first of November" (*WF* 83). Thoreau favors the late season for the wild fruits, those left behind by farmers and merchants. The owner of old trees "has not faith enough to look under their boughs," but the gleaner's "faith is rewarded by finding the ground strewn with spirited fruit,—some of it, perhaps, collected at squirrel-holes, with the marks of their teeth by which they carried them,—some containing a cricket or two silently feeding within, and some, especially in damp days, a shelless snail" (85). The late season apples "are choicest fruit to the walker. But it is remarkable that the wild apple, which I praise as so spirited and racy when eaten in the fields or woods, being brought into the house, has frequently a harsh and crabbed taste. The Saunterer's Apple not even the saunterer can eat in the house" (85). The reason for this placing is not hard to see and taste: "These apples have hung in the wind and frost and rain till they have absorbed the qualities of the weather or season, and thus are highly *seasoned,* and they *pierce* and *sting* and *permeate* us with their spirit. They must be eaten in *season,* accordingly,—that is, out-of-doors" (86). It is tempting to read the italicized words in this brief passage as signals for emphasis in the lecture halls of Concord or Bedford. One imagines an audience listening quietly and then being aroused by the string of five active words, all marked by the /i/ sound. It would mark a season in the winter lyceum program. The audience might feel the words "pierce" and "sting" and "permeate" them with their spirit.

It is never a large step from literal to figurative fruits. Thus, Thoreau concludes, "there is one thought for the field, another for the house. I would have my thoughts, like wild apples, to be food for walkers and will not warrant them to be palatable if tasted in the house" (*WF* 87). As "Wild Apples" proceeds, the individual walker becomes a party of walkers and perhaps even a community of walkers who are also the readers of Thoreau's work. The motif and its evolving role in "Wild Apples" support Lance Newman's view of *Wild Fruits* as evoking a utopian, communitarian vision.[4] Perhaps even stronger support comes in the list of names for wild apples. Tongue in cheek, as usual, Thoreau gives some Latin names "for the benefit of those who live where English is not spoken—for they are likely to have a world-wide reputation" (*WF* 89). He repeats the name "the Saunterer's Apple" and notes that "you must lose yourself before you can find the way to that" (89). Perhaps the closest analog to this saying occurs in a passage in which Jesus foretells his death and warns his followers of the cost of believing in him: "And when he had called the people unto him with his disciples also, he said unto them, Whosoever will come after me, let him deny himself, and take up his cross, and follow me. For whosoever will save his life shall lose it; but whosoever shall lose his life for my sake and the gospel's, the same shall save it. For what shall it profit a man, if he shall gain the whole world, and lose his own soul? Or what shall a man give in exchange for his soul?" (Mark 8:34–37).[5] A second name, "our Particular Apple, not to be found in any catalogue, *Malus pedestrium-solatium*," is at least as worshipful. The Latin translates as the Walker's Solace Apple, or Walker's Comfort Apple. Indeed, *pedestrium* might best be translated as "infantry" or "foot soldiers," and in conjunction with "saunterer" and the etymology for the word that Thoreau proposes in "Walking" (*Exc* 186), we may think of the walkers as bound in some fashion for the holy land, begging for wild apples along the way.

Thoreau ends the "Wild Apples" entry by echoing the language of the longer introduction to *Wild Fruits,* contrasting the "frozen-thawed" apple with the "imported half-ripe fruits of the torrid South" (*WF* 91). Then he announces a prophecy that seems meant to be taken seriously: "The era of the wild apple will soon be past. It is a fruit which will probably become extinct in New England" (91). The walker of the future will likewise suffer: "I fear that he who walks over these fields a century hence will not know the pleasure of knocking off wild apples. Ah, poor man, there are many pleasures which he will not know" (92). Thoreau prophesies the loss of the wild itself, and then he joins it with the prophet Joel's jeremiad that ends the published essay: "The apple-tree, even all the trees of the field, are withered: because joy is withered away from the sons of men" (*Exc* 289).[6]

THE MOST PERSEVERING NATIVE AMERICANS

If the conclusion to the "Wild Apples" essay were a synecdoche, we would read *Wild Fruits* as a dark prophecy, an angry voice out of the whirlwind. And there are strong elements of the Old Testament prophet in Thoreau's late work, as the John Brown papers, "Slavery in Massachusetts," and "Life without Principle" clearly show. In "Life without Principle" in particular, Thoreau develops the theme of "getting a living" by giving the reader "a strong dose of myself," perhaps the most unadulterated dose we could take.[7] But if we take that strong dose along with the other three lectures published posthumously in the *Atlantic Monthly*—"Autumnal Tints," "Walking," and "Wild Apples"—we may begin to appreciate the variety of tones Thoreau is capable of producing, both in the lyceum and in print. We also may begin to hear the underlying social critique and utopian communitarianism that mark *Wild Fruits*.

One productive way of testing this reading of *Wild Fruits* comes in the longest entry of the book, "Black huckleberry."[8] It is less polished than "Wild Apples," but that is in its favor. Thoreau permits himself to praise the many varieties of huckleberry and describe one "firmer berry" as "the most marketable" of the whortleberries (*WF* 38), a financial superlative one would never find in "Wild Apples."[9] "Black huckleberry" presents a familiar plant to an audience already acquainted with the fruit: "This, as you know, is an upright shrub . . . [that] abounds over but a small part of this area, and there are large tracts where it is not found at all" (37). The familiarity of the plant and its relationship to the audience create common ground in the discussion of the fruit and its values.

In a draft introduction to the lecture, Thoreau strikes yet more deeply at the familiar relations:

> I presume that every one of my audience knows what a huckleberry is,—has seen a huckleberry, gathered a huckleberry, and, finally, has tasted a huckleberry,—and, that being the case, I think that I need offer no apology if I make huckleberries my theme this evening.
>
> What more encouraging sight at the end of a long ramble than the endless successive patches of green bushes,—perhaps in some rocky pasture,—fairly blackened with the profusion of fresh and glossy berries?
>
> There are so many of these berries in their season that most do not perceive that birds and quadrupeds make any use of them, since they are not felt to rob us; yet they are more important to them than to us. We do not notice the robin when it plucks a berry, as when it visits our favorite cherry tree, and the fox pays his visits to the field when we are not there. (*J* XIV:310)

The familiar relationships proliferate in several ways here. The speaker and the audience are on familiar, easy terms; the huckleberry bushes provide an "encouraging sight" to tired ramblers; and the profuse, wild berries are important for birds and other animals, without human beings incurring any loss from their visits.

The familiar tone extends to Thoreau's personal experiences in Concord and its neighborhood. For instance, to illustrate the variety he calls "the red huckleberry, the *white* of some (for the less ripe are whitish)," he tells the entertaining story of surveying a farm and being paid with "a quart of red huckleberries," a gesture he finds "ominous," and one that teaches him to "beware of red-huckleberry gifts in future" (WF 39). In another passage, he recalls how he, "a lad of ten, was despatched to a neighboring hill alone," to pick enough for a huckleberry pudding to entertain an itinerant "mantua-maker" (55). The conscientious boy was sure to pick enough by eleven o'clock—"and all ripe ones, too, though I turned some round three times to be sure they were not premature. My rule in such cases was never to eat one till my dish was full, for going a-berrying implies more things than eating the berries. They at home got nothing but the pudding, a comparatively heavy affair, but I got the forenoon out of doors—to say nothing about the appetite for the pudding" (55).

The humor in these anecdotes is gently charming, conveying a sense of pleasure in berry-picking and in storytelling. Another pleasure comes in discovery. In a passage taken from the Journal of 4 August 1856 (*J* VIII:444–45), Thoreau tells of finding, "on the side of Conantum Hill," huckleberries, blackberries, and blueberries "literally five or six species deep." He details the species in order to illustrate the "unusual profusion" of the wild fruits, and the details allow him to show how he guides the huckleberry parties to their best experiences: "You went daintily wading through this thicket, picking perhaps only the finest of the high blackberries, as big as the end of your thumb, however big that may be, or clutching here and there a handful of huckleberries for variety, but never suspecting the delicious, cool, blue-bloomed ones, which you were crushing with your feet under all. I have in such a case spread aside the bushes and revealed the last kind to those who had never in all their lives seen or heard of it before" (WF 53). Here Thoreau acts as a guide both to his berry-picking companions and to his listeners and readers.

The multiple guide instructs us in social modes of being and knowing, and the guide stems directly from Thoreau's botanical excursions, recorded in the Journal for over a decade. In such berry-picking, the wild fruits "seem offered to us not so much for food as for sociality, inviting us to a picnic with Nature. We pluck and eat in remembrance of her. It is a sort of sacrament, a communion—the *not* forbidden fruits, which no serpent tempts us to eat. Slight and innocent savors which relate us to Nature, make us her guests, and entitle us to her regard and protection" (*WF* 52). The echo of the Last Supper scene most directly evokes "this do in remembrance of me" (Luke 22:19), and the "sort of sacrament, a communion" seems only mildly irreverent. Such profuse huckleberry and blueberry bushes make Thoreau "think of them as fruits fit to grow on the most Olympian or heaven-pointing hills," and "you eat these berries in the dry pastures where they grow not to gratify an appetite, but as simply and naturally as thoughts come into your mind, as if they were the food of thought, dry as itself, and surely they nourish the brain there" (*WF* 52–53).

As in "Wild Apples," Thoreau steps easily from literal to figurative berries. The profusion of berries leads to the profusion of symbolic, "heaven-pointing" language. Thus the communion of berry-pickers resonates with the guide's literary wish: "I would have my thoughts, like wild apples, to be food for walkers and will not warrant them to be palatable if tasted in the house" (*WF* 87). Even more prominently than in "Wild Apples," the berry-picking parties become a figure for community education in the fields, swamps, and woods. In a charming aside, Thoreau notes that "to the old walker the straggling party itself, half concealed amid the bushes, is the most novel and interesting feature" (55). The guide and his party "may be nicknamed 'huckleberry people,'" for the emissaries from large towns and cities "come more for our berries than they do for our salvation" (55). The saving of the huckleberry people comes finally from the "sense of freedom and spirit of adventure," and for the guide "an expansion of all my being," which he would not exchange "for all the learning in the world. Liberation and enlargement—such is the fruit which all culture aims to secure" (57).

Like "Wild Apples," the "Black huckleberry" entry ends with tones of the jeremiad, but the prophecy of evil days does not seem so darkly foreboding. Indeed, it may partake of liberation and enlargement. The "true value of country life" is not to be found in the market. The local berry parties persist, so that Thoreau looks to England and Europe to predict that "the wild fruits of the earth disappear before civilization, or only the husks of them are to be found in large markets" (*WF* 57). The modern tendency toward division of labor subverts the "ultimate fruit of the huckleberry field," which we find in picking the fruits ourselves. All efforts to commercialize and privatize and divide run counter to "the spirit of the huckleberry" (58), and by gathering berries in our fields we are "gathering health and happiness and inspiration and a hundred other far finer and nobler fruits than berries which are found there." Such fruits will never be brought to market, and the market is thus striking "one more blow at a simple and wholesome relation to nature" (58). The jeremiad echoes the introduction to *Wild Fruits* and, interestingly, counters Thoreau's ideas of getting a living by marketing huckleberries, both suggested and rejected in *Walden* (W 69–70, 173). Thoreau is always representing what he calls "the huckleberry interest," in opposition to the other interests his audience may bring with them: "Every interest, as the codfish and the mackerel, gets represented but the huckleberry interest. The first discoverers and explorers of the land make report of this fruit, but the last make comparatively little account of them" (*J* XIV:282). In this entry from the 29 November 1860 Journal, the contrast between the "first discoverers and explorers of the land" and "the last" seems to oppose past and present, though it is unclear whether he means the white "discoverers and explorers" or the Native discoverers and explorers who preceded them.

For modern readers, the most important way in which "Black huckleberry" differs from "Wild Apples" is Thoreau's treatment of Native Americans. In "Wild Apples," he presents himself as "an outdoor man" (WF 87), "semi-civilized" (91), and his preferred fruits are "backwoodsmen" (80). While he affirms that "it takes a savage or wild taste to appreciate a wild fruit" (87), he also states confidently that the indigenous crab apple is no "hardier than those backwoodsmen among the apple trees, which, though descended from cultivated stocks, plant themselves in distant fields and forests, where the soil is favorable to them. I know of no trees which have more difficulties to contend with, and which more sturdily resist their foes. These are the ones whose story we have to tell" (80). He characterizes the blueberry and huckleberry bushes, by contrast, as "the most persevering Native Americans, ready to shoot up into place and power at the next election among the plants, ready to reclothe the hills when man has laid them bare and feed all kind of pensioners" (44). The figuration moves from Native, Indigenous Americans to the colonizing world of elections and pensioners, and Thoreau consistently considers this an historical displacement as well. But in "Black huckleberry" he rails against the scientific name *Gaylussacia resinosa* because the genus is named after a French chemist: "What if a committee of Parisian naturalists had been appointed to break this important news to an Indian maiden who had just filled her basket on the shore of

Lake Huron! It is as if we should hear that the daguerreotype had been *finally* named after the distinguished Chippeway conjurer, The-Wind-that-Blows" (37).

Thoreau directly relates the huckleberries to Native Americans in the fundamental sense of geographical distribution. He notes that the huckle-berry and other native fruits "have a range, most of them, very nearly coterminous with what has been called the Algonquin Family of Indians, whose territories embraced what are now the Eastern, Middle, and North-western States, and the Canadas, and surrounded those of the Iroquois in what is now New York. These were the small fruits of the Algonquin and Iroquois Families" (WF 46). The past tense shows that Thoreau considers the Algonquin and Iroquois as former inhabitants of the States, a view that expresses his dominant white perspective. But the settler perspective is not monolithic in the entry. The Native Americans are important teachers for the colonizers: "They taught us not only the use of corn and how to plant it, but also of whortleberries and how to dry them for winter. We should have hesitated long before we tasted some kinds if they had not set us the example, knowing by old experience that they were not only harmless but salutary. I have added a few to my number of edible berries by walking behind an Indian in Maine and observing that he ate some which I never thought of tasting before" (46). The last line refers to the 1857 excursion with Joseph Polis, which represents Thoreau's most enlightened and en-larged encounter with Native Americans. Thoreau's perspective is mixed, combining the mainstream white settler view of Native Americans as "our" past teachers and the writer's recent personal experience of walking behind a living Penobscot guide.

As "Black huckleberry" proceeds, Thoreau amasses the testimonies of his white "authorities," a practice he uses in *Cape Cod* and *The Maine Woods* to emphasize the so-called discoveries of French and British colonial explorers and settlers. These would usually be the "first discoverers and explorers" Thoreau mentions in the Journal passage concerning the "huckleberry interest." Here, however, he does so to show that the Native Americans "did not learn the use of these berries from us whites" (WF 46). Even if his sources suggest otherwise, he insists that "it was the whites who imitated the Indians rather" (48). As he delivers the summary of white explorers and botanists from the seventeenth century to the present, he brings the history of Native Americans to the present as well: "Hence you see that the Indians, from time immemorial down to the present day, all over the northern part of America, have made far more extensive use of the whortleberry at all and in various ways than we, and that they were far more important to them than to us" (49). The mixed perspective is still evident here, and Thoreau's enlightened view of the Indigenous inhabitants, "from time immemorial down to the present day," is partly undercut by the simple past tense of the final clause.

In a final suggestion concerning the present role of Native Americans, Thoreau recurs to the genus name for the huckleberry and imagines that Native American names should be used by botanists: "I think that it would be well if the Indian names were as far as possible restored and applied to the numerous species of huckleberries by our botanists, instead of the very inadequate Greek and Latin or English ones at present used. They might serve both a scientific and popular use. Certainly it is not the best point of view to look at this peculiarly American family, as it were, from the other side of the Atlantic. It is still in doubt whether the Latin word for the genus *Vaccinium* means a berry or a flower" (WF 50). No modern reader would seriously interpret Thoreau's ideas as progressive, but there is no doubt that he sees values in Native American culture, knowledge, and language, values that should be acknowledged by scientists and the general public. In the context of the other arguments for the "huckleberry interest," it is no exaggeration to say that Thoreau is trying to represent the Native American interest as well. Both the Native Americans and the huckleberries are a "peculiarly American family."

The same can be said of Thoreau's view of huckleberries in general. In a passage derived from the Journal of 28 May 1854 (*PJ* 8:161), he praises the flowers of the "*Vaccinieae* or Whortleberry Family," taking them as a promising sign of new fruits: "This crop grows wild all over the country—wholesome, bountiful, and free, a real ambrosia. And yet men, the foolish demons that they are, devote themselves to the culture of tobacco, inventing slavery and a thousand other curses for that purpose, with infinite pains and inhumanity go raise tobacco all their lives, and that is the staple instead of huckleberries. Wreaths of tobacco smoke go up from this land, the only incense which its inhabitants burn in honor of their gods" (WF 51–52). The new fruits, in the largest sense of the word, would come from a cultural reform, a turn away from the likes of tobacco and its agricultural, social, economic, and physical curses, most especially the curse of slavery. It could entail a turn to the values and uses white settlers first learned from Native American cultures, a return to the wild fruits of the land. Thoreau is as serious and high minded here as he is in a later passage, in which he describes, with real horror, schoolboys reenacting the policies of the Young America movement of the 1840s. The children would lay claim to berry patches, shouting, "I speak for this place." Sometimes, Thoreau remarks sadly, that was "considered good law for the huckleberry field. At any rate, it is a law similar to this by which we have taken possession of the territory of Indians and Mexicans" (56).

A final important piece of evidence for Thoreau's representations of the huckleberry interest and its allies comes in the very beginning of *Wild Fruits*. He remarks, "As I sail the unexplored sea of Concord, many a dell and swamp and wooded hill is my Ceram and Amboyna" (*WF* 3). The sentence appears in the 23 November 1860 Journal, as an isolated paragraph (*J* XIV:262). We see the ironic contrast of the familiar Concord and the exotic islands of the East Indies. Of course, the irony cuts in favor of the familiar, just as the introduction to *Wild Fruits* favors the unexplored "native fields" and the local berries one has not yet learned to name (*WF* 3). In the Journal, a paragraph follows that Thoreau does not use in all the pages of *Wild Fruits*. And yet it is the most telling passage for the liberation and enlargement inherent in the "huckleberry interest":

> At first, perchance, there would be an abundant crop of rank garden weeds and grasses in the cultivated land,—and rankest of all in the cellar-holes,—and of pinweed, hardhack, sumach, blackberry, thimble-berry, raspberry, etc., in the fields and pastures. Elm, ash, maples, etc., would grow vigorously along old garden limits and main streets. Garden weeds and grasses would soon disappear. Huckleberry and blueberry bushes, lambkill, hazel, sweet-fern, barberry, elder, also shad-bush, choke-berry, andromeda, and thorns, etc., would rapidly prevail in the deserted pastures. At the same time the wild cherries, birch, poplar, willows, checkerberry would re-establish themselves. Finally the pines, hemlock, spruce, larch, shrub oak, oaks, chestnut, beech, and walnuts would occupy the site of Concord once more. The apple and perhaps all exotic trees and shrubs and a great part of the indigenous ones named above would have disappeared, and the laurel and yew would to some extent be an underwood here, and perchance the red man once more thread his way through the mossy, swamp-like, primitive wood. (*J* XIV:262–63)

Thoreau imagines and enacts an elaborate fantasy of postapocalyptic succession in this passage. The crops of cultivated land are gone already when he begins the narrative. The white settlers' homes are only "cellar-holes."

Weeds and grasses take over, then the briars and ground cover plants. Fast-growing trees like elm and ash spring up in the edges; indigenous plants like blueberry and huckleberry and the rest take over the deserted pastures. The final sentences create the image of a restored native temperate broadleaf and conifer forest, with an intact structure of shade-tolerant understory plants. The shrubs, small trees, and tall mature trees take over and "occupy the site of Concord once more." In the very last image, moreover, we see the surprising reversal in Thoreau's vision of forest succession and a new response to the mismanagements recorded in *Dispersion of Seeds*. There are no white settlers in these woods, no "first discoverers and explorers of the land." Instead, our guide imagines "perchance the red man once more thread his way through the mossy, swamp-like, primitive wood." In Thoreau's vision, the first are once more the first. The power and vitality of the botanical world yield the ultimate liberation and enlargement, the fruits of inhabitation and restoration.

SMALL FRUITS AND LARGE QUESTIONS

Many of the longer entries in *Wild Fruits* focus on what are called "small fruit," a North American mass noun or plural noun, meaning "the edible fruit of any kind of low-growing perennial, fruit-bearing plant" (*OED*). The term dates to the eighteenth century and still enjoys currency in horticultural catalogs and websites. The longer small fruit entries in *Wild Fruits* include "Strawberry" (10–17), "Early low blueberry" (21–26), "High blueberry" (30–36), "Common cranberry" (102–7), "Barberry" (139–43), "Wild grape" (150–57), and "European cranberry" (164–70). Most of these entries are wholly from Thoreau's pen, but others are the editorial reconstructions of Bradley Dean. For instance, the entry on "European cranberry" (*WF* 164–70) includes four passages that Dean incorporates into the text according to an order indicated by Thoreau in the "Notes on fruits" manuscript. Three of the four come from the Journal of 30 August, 2 September, and 3 September 1856, while the final quotation from the British horticulturalist John Loudon comes from Thoreau's Common Place Book (*WF* 344). In fact, most of the "European cranberry" entry comes from the 30 August 1856 Journal, an extended account of an afternoon of "a-cranberrying" in the sphagnum bog Thoreau names the Vaccinium Oxycoccus Swamp (*J* IX:35). As we saw in an earlier reading of this Journal entry, the excursion begins with a comparison of the small cranberry and common cranberry, but it leads Thoreau to a larger, more expansive sense of discovery. The Journal entry runs over ten pages (IX:35–46), but Dean leaves out a three-page section that he likely considered a digression about the discovery of a new kind of hairy huckleberry, *Gaylusaccia Dumosa* var. *hirtella* (IX:41). That new variety of huckleberry leads to the writer's exclamation, "I seemed to have reached a new world, so wild a place that the very huckleberries grew hairy and were inedible" (IX:42), and to a meditation on local wildness (IX:42–44). The strictures of the *Wild Fruits* format provide a reasonable editorial decision, but the omitted passage from the Journal is a significant part of Thoreau's botany.

Dean's editorial reconstruction of the "European cranberry" entry teaches more than one lesson. The idea of wildness does not disappear from the *Wild Fruits* entry with the omitted pages of the digression, and the language of ecstatic worship remains as well. When Thoreau comments that "the more thrilling, wonderful, divine objects I behold in a day, the more expanded and immortal I become" (*WF* 169), however, he is in fact referring both to the hairy huckleberry and to two varieties of the European cranberry, in addition to colonies of ants and the sphagnum mountains he sees them inhabit (*J* IX:43). The unfinished *Wild Fruits* contains, in a strange way, material that does not appear in print, not even in the miraculous reconstructions of Bradley Dean.

The example of "European cranberry" opens *Wild Fruits* in ways that the long, relatively polished entries like "Wild Apples" and "Black huckleberry" do not. Rather than read those two entries as exclusive models for all the entries of an unfinished project, we can read the long, short, and note-like entries as three different modes of knowing and writing, not just as stages of an imagined drafting process that would run in a linear direction. Thus, for instance, it is instructive to read "European cranberry" in conjunction with "Common cranberry" (*WF* 102–7). The latter entry compiles and revises Journal passages (especially 20–21 November 1853) to give a series of anecdotes about raking cranberries. Over a period of several days in November 1853, Thoreau discovers a host of them on the vines, drifting at the bottom of flooded meadows. He finds a broken rake up the river and uses it to harvest a boatload of berries, which he cleans and then sells in Boston. The work of "raking so many cranberries out of the water made me quite conversant with the materials which compose the river wrack" (104). Somewhat surprisingly, he also determines the best way to harvest the berries for potential markets. The failure of *A Week on the Concord and Merrimack Rivers,* and the $100 debt to his publisher, leads Thoreau to consider "speculating in cranberries," but he finds that the price he would pay in New York City is less than the price he would receive in Boston (104–5). With that new knowledge of markets, he sticks to selling pencils.[10]

The entry in *Wild Fruits* is a less detailed, more humorously wry look at such speculations in small fruits. As a way of marking the follies of business, Thoreau follows with an account of how, at the age of twelve, he was chased off a cranberry pasture by one "Old Foster," who pursued him even into the village. "I did not know till then," he claims, "that cranberries were private property" (*WF* 105). As a final stage of knowledge, he states that cranberries deserve to be eaten in the open air of spring: "They are never so beautiful as in water. In the markets and aboard ships they ask if you will have wet or dry, if they shall be transported on deck or in the hold, but to my mind the only wet cask in which to get them is the flooded meadows in the spring" (106).

The stages of knowledge in the "Common cranberry" entry sharpen the reflections on "enterprise" in the "European cranberry" entry. Questioning his project of gathering the "small cranberry," Thoreau realizes that "it is these comparatively cheap and private expeditions that substantiate our existence and batten our lives—as, where a vine touches the earth in its undulating course, it puts forth roots and thickens its stock" (*WF* 165–66). The expedition to the Vaccinium Oxycoccus Swamp is one of "carrying out deliberately and faithfully the hundred little purposes which every man's genius must have suggested to him," and it yields "such a sauce as no wealth can buy" (166). As if recalling his earlier attempts with the common cranberry, Thoreau describes the cranberry swamp of August–September 1856 as "yielding its crop to me alone," which he does not regard "in the light of their pecuniary value" (167). The excursion in the swamp leads him into new territory, "wild as a square rod on the moon," and he sees the two varieties of small cranberries as "meteoric, aerolitic," and worthy of reverence (168). By reading across entries and consulting the sources in the Journal, we enrich the value of *Wild Fruits* and deepen the significance of Thoreau's botany.

Longer entries like "Early low blueberry" (*WF* 21–26) and "High blueberry" (30–36) yield further insights when read together. First, they show that Thoreau is a persistent student of varieties. He cites "earlier European botanists" as distinguishing two varieties of the early low blueberry or dwarf blueberry, *Vaccinium Pennsylvanicum* and *Vaccinium Canadense,* though these are treated by Asa Gray as separate species (Gray 312). Thoreau's accounts of botanical excursions around Concord, on Monadnock, in other New Hampshire mountains, and in Maine lead him to accept the geographical and physical distinction between the two species, even though he recognizes their overlap on mountaintops and rocky summits (WF 24). Indeed, the entry concludes by noting two more "allied kinds" in the White Mountains, "the bog bilberry (*Vaccinium uliginosum*) and dwarf bilberry (*Vaccinium caespitorum*)" (26). In the "High blueberry" entry, Thoreau notes that the two common varieties are "the blue and the black (*Vaccinium corymbosum* and its variety, *atrocarpum*)" (30). In effect, the entries provide a kind of ground-truthing of botanical manuals, and they open the discussion from an ostensible focus on one species to a panorama of many related species and varieties.

Ground-truthing can arise from vocabulary. For instance, another name for the common blueberry is the swamp blueberry, and Thoreau finds the fruit most often in swamps and bogs, and along the shores of ponds. In the "High blueberry" entry, he narrates at least six separate excursions to neighboring blueberry preserves, each one adding to the description and developing images of "inexhaustible abundance" through the summer (WF 32). The descriptions in fact move from the berries to the bushes themselves, which become the "venerable" focus of the entry as the berry-picking season progresses. The bushes are most visible for inspection in winter, and then they show themselves as "scraggy, gray, dead-looking," and "covered with lichens, commonly crooked, zigzag, and intertwisted with their neighbors" (34). At Goose Pond, Thoreau counts forty-two annual rings on one shrub and measures one trunk as "eight and a half inches in circumference at the butt" (34). The passage in *Wild Fruits* joins two excursions to the pond, in December 1857 and February 1858 (*J* X:228, 278), and the entry represents knowledge and expertise accumulated over several years and seasons.

As the "High blueberry" entry moves toward a conclusion, Thoreau notes "the largest and handsomest" blueberry shrub, encountered on Sassafras Island in Flint's Pond, during excursions on 22 and 24 December 1859 (*J* XIII:39–40, 46–47). The shrub is actually a "small tree or clump of trees, about ten feet high and spreading the same or more, and is perfectly sound and vigorous." The bark is reddish, "at intervals handsomely clothed with large yellow and gray lichens," and the tree appears to be "about sixty years old" (WF 35). Thoreau sits in the tree, some four feet from the ground, and reflects that "perhaps yet larger ones were to be seen here before the whites came to cut down the woods. They are often older than many whole orchards of cultivated fruit trees and may have borne fruit before the writer was born" (36). The writing incorporates physical movement, a fresh location from which to view the plant and describe it, and a new perspective on age that suggests a connection to human life spans. These are fruits of plant-thinking, discovered through repeated botanical excursions and years of faithful writing.

Immediately after this entry on the "High blueberry," we find a short, four-paragraph description of the "Late low blueberry" (*WF* 36–37). The entry reads as if it belongs with the previous one, beginning with "About the same time." It describes another common species, *Vaccinium vacillans,* a firm berry that Thoreau distinguishes from the early low blueberry as "more like solid food, hard and bread-like, though at the same time more earthy" (36). Though brief, the description takes us from first fruiting to the middle of September, when the writer finds the fruits "quite sound, in fact, after all the rest of the plant has turned to a deep crimson, which is its autumnal tint. These almost spicy, lingering clusters of blueberries contrast strangely with the bright leaves" (37). This apparently complete description suggests that short entries could be interspersed with long ones, and that some of the long entries could merge productively with short ones treating related species. This idea is yet another way in which the edited *Wild Fruits* creates imaginative possibilities for the book Thoreau could not live to finish.

As encouraging as such a reading may be, it does not always emerge from the juxtaposition of long and short entries. We can see the less productive side by comparing the late low blueberry to "Red and fetid currants" (*WF* 59), the entry immediately following "Black huckleberry" (37–59). The entry on currants is made up of two brief Journal notes on the first fruiting ("say July third"), a short paragraph on unnamed "old writers of New England," and a five-sentence paragraph on Thoreau's encounters with "the fetid currant (*Ribes prostratum*)" by the roadside in New Hampshire, on Monadnock, and in the White Mountains (59). The entry on "Wild gooseberry," another species of *Ribes,* is a similarly rough sketch (62–63). In no way can a sympathetic reading produce an outline of coherence in these neighboring entries. Small fruits sometimes remain small.

Among the small fruits, "Strawberry" (*WF* 10–17) is an early, relatively polished entry. Thoreau notes that it is "our first edible fruit to ripen," before the cultivated kinds begin to redden. The wild strawberry, like the wild apple, is a discovery for the walker, who must look "in the most favorable places," such as "the southerly slope of some dry and bare hill" (11). Like the apple, the strawberry is remarkable for its fragrance, "that sweet scent of the earth of which the ancients speak" (12). It requires "a sort of Indian knowledge, acquired by secret tradition," to find the early wild strawberries, and again like the wild apple, the wild fruit outshines the cultivated garden and market varieties (13). The wild strawberry marks the arrival of "the season of *small fruits*" (13), and to the writer that suggests a gap between "our hopes and their fulfillment" (14). The walker must be contented with two or three handfuls a year, and only farther north, in New Hampshire and Maine, does the wild strawberry attain real profusion (15). In rarity and secrecy, then, the wild strawberry resembles the huckleberry. Thoreau surveys Indian names for the strawberry and ultimately rejects the English word: "Better call it by the Indian name of heart-berry, for it is indeed a crimson heart which we eat at the beginning of summer to make us brave for all the rest of the year, as Nature is" (17). In form, length, and coherence, the "Strawberry" entry represents a kind of model essay.

Somewhat less finished is "Barberry" (*WF* 139–43), which begins with a summary of berrying dates from 1852 to 1859. Thoreau quickly settles into one of the characteristic patterns of the *Wild Fruits* entries, noting the season for berrying, describing the bush and its ripe fruit, and then naming some of his favorite "resorts" (140). He recounts a barberry party with his aunts and his sister Sophia, then expands on his experience as a "dexterous barberry picker" (141). The entry does not develop its descriptions by detailing specific excursions or events, nor does Thoreau use his historical sources to much advantage. He includes Jones Very's sonnet "The Barberry Bush," but the love poem adds only a rather obvious analogy to the image of the fruit and its meanings.

The "Barberry" entry is arguably more developed than "Wild Grape" (*WF* 150–57), "Acorns" (178–86), "Chestnut" (209–16), "Walnuts" (216–20), and "Pitch pine" (227–32), all of which remain on the level of dated Journal compilations. In reading these entries, we see that the rhetorical or meditative structure of an essay, or the narrative spine of an excursion, most often gives an entry the quality of finish and transforms the entry into a deeper treatment of fruits. For those reasons, a short entry like "Hazel" (160–62) appears very nearly complete, and an entry on milkweed, "*Asclepias cornute*" (196–98), reaches heightened meaning in a question like this: "Who could believe in prophecies of Daniel or of Miller that the world would end this summer, while one milkweed with faith matured its seeds?" (198).

PERCEPTIONS AND VALUES

The concluding section of Wild Fruits consists of paragraphs drawn from Journal entries, dating from as early as 23 August 1853 and as late as 3 January 1861. There may be some uncertainty about the order of these paragraphs and their role as a conclusion for the entire draft of Wild Fruits, but Bradley Dean follows a manuscript in the Berg Collection to publish the section, and it is certainly safest to follow his lead.[11] The beginning of the section strikes the major chord of the piece: "How little we insist on truly grand and beautiful natural features. There may be the most beautiful landscapes in the world within a dozen miles of us, for aught we know—for their inhabitants do not value nor perceive them, and so have not made them known to others" (WF 233). The nearly two hundred entries of the book reveal a host of wild fruits in the local neighborhood and detail thousands of excursions to the rivers, swamps, bogs, meadows, pastures, and forests that Thoreau made over the final decade of his life. These are the beautiful landscapes within a dozen miles of us, unvalued and unperceived by many local inhabitants. The purpose of Wild Fruits is to make these features known, and to show that they are "truly grand and beautiful." In doing so, Thoreau would also educate his neighbors in the perception of value and the value of perception.

The overarching state of New England towns is to ignore "in what its true wealth consists" (WF 233). The town of Boxborough, eight miles west of Concord, provides an example that resonates in both Dispersion of Seeds and Wild Fruits, for it holds a "noble oak wood" that is "the handsomest and most memorable thing" Thoreau has seen there. Local histories never mention the oaks, and for that they earn the writer's scorn: "Wherever men have lived, there is a story to be told, and it depends chiefly on the story-teller or historian whether that is interesting or not" (235). Boxborough has preserved the oaks, but only because the land yields higher tax dividends than it had done previously. For these reasons, Thoreau proposes that the state of Massachusetts "purchase and preserve a few such forests." In this idea, Thoreau clearly echoes the third chapter of The Dispersion of Seeds and the coda to "Chesuncook," in which he calls for "national preserves" of forest land (MW 156).

A second theme of the conclusion is the true value of common property to a community. This is the "true fruit of Nature," which cannot be "bribed by any earthly reward. No hired man can help us to gather that crop" (*WF* 235). According to Thoreau, the Native Americans held the fruits of nature in common; thus they are far superior to "civilized men," and how they "have arranged it." The natural features of the land, which embody the values of the community, "have a high use which dollars and cents never represent" (236). Thoreau names three particular natural features as deserving protection. In Concord, he finds the river the great "ornament and treasure," but the town has done "nothing to preserve its natural beauty" (236). Likewise, "any central and commanding hilltop" or "mountaintop" in a township should be preserved as a natural "temple" (237). The third possibility is to set aside "a park, or rather a primitive forest, of five hundred or a thousand acres, either in one body or several, where a stick should never be cut for fuel, nor for the navy, nor to make wagons, but stand and decay for higher uses—a common possession forever, for instruction and recreation" (238). Though Thoreau relegates Native Americans to a past ideal of community, his three suggestions for preservation restore the ideal in the present, or in an imagined future. We recognize the suggestions from other excursions and other essays; indeed, the call for preservation is a vital part of Thoreau's plant-thinking, a way of putting human and more-than-human realms into a new relation.

Thoreau closes *Wild Fruits* with a peroration, admonishing his readers to "live in each season as it passes" (*WF* 238). The idea hearkens back to the Journal of 23 August 1853 and to the time in which Thoreau was himself learning to live *in* each season. The seasonal, phenological order of *Wild Fruits* stands as an object lesson in how to pay attention to the wild plants around us, how to discover their highest uses. By developing our senses, opening "all your pores" to the "tides of Nature, in all her streams and oceans, at all seasons," we bring our powers of perception into real health, for "miasma and infection are from within, not without." The figuration stresses the interior transformation of readers, and the transformation is a kind of baptism within the "tides of Nature." The fluid imagery bespeaks perpetual motion and change, moreover, the vital force of *phusis*. In another figure, the transformation is a form of communion. Most of all, Thoreau tells us, take nourishment from the wines bottled by nature, kept "in the skin of a myriad fair berries" (239). Like a botanical eucharist, the berries offer the sustenance of spirit and body, for nature is "but another name for health." Our perceptions are valuable in so far as they delve into nature and dwell within nature. Then we will be *"well"* in the depths of our sensations and we will live "fairly *in* those seasons" (239).

Emerson concludes his famous eulogy for Thoreau with a pensive version of plant-thinking, at least of an Emersonian kind. A plant leads him to his peroration, unlike Thoreau's conclusion to *Wild Fruits:*

> There is a flower known to botanists, one of the same genus with our summer plant called " Life-Everlasting," a *Gnaphalium* like that, which grows on the most inaccessible cliffs of the Tyrolese mountains, where the chamois dare hardly venture, and which the hunter, tempted by its beauty, and by his love, (for it is immensely valued by the Swiss maidens,) climbs the cliffs to gather, and is sometimes found dead at the foot, with the flower in his hand. It is called by botanists the *Gnaphalium leontopodium,* but by the Swiss *Edelweisse,* which signifies *Noble Purity.* Thoreau seemed to me living in the hope to gather this plant, which belonged to him of right. The scale on which his studies proceeded was so large as to require longevity, and we were the less prepared for his sudden disappearance. The country knows not yet, or in the least part, how great a son it has lost. It seems an injury that he should leave in the midst his broken task, which none else can finish, —a kind of indignity to so noble a soul, that it should depart out of Nature before yet he has been really shown to his peers for what he is. But he, at least, is content. His soul was made for the noblest society; he had in a short life exhausted the capabilities of this world; wherever there is knowledge, wherever there is virtue, wherever there is beauty, he will find a home.[12]

The eulogy is full of great praise for Thoreau, but for many readers the "broken task" of the final decade has cast a shadow over his life and work. The unfortunate word is "broken," implying a sense of a misbegotten goal or mistaken purpose. It may also imply a sense that Thoreau was not *well,* and that he did not live "fairly *in* those seasons." But if this study of the Journal, the Kalendar project, *The Dispersion of Seeds,* and *Wild Fruits* is persuasive, it shows that Thoreau's large-scale botanical work of the last decade was clearly unbroken. Thoreau's botany is ultimately an unbroken task, even if we are still completing it in our present studies.

Walking in the Anthropocene

If Michael Marder's *Plant-Thinking* is a kind of philosophical prolegomenon to critical plant studies, Thoreau's botany provides a set of concrete, interdisciplinary practices for encountering plants and finding meaning in our relationship to their lives. In *Walden Warming* (2014) and numerous articles, Richard Primack has shown that Thoreau and other citizen scientists of Concord recorded faithfully the phenological data that help us see climate change at work in our world. This is important science, not least for showing new uses for reading Thoreau's Journal. As the Primack Lab website shows, too, the intersections of science and the humanities can open new possibilities for reckoning with climate change in our present world.[1] How we translate the excursions into present-day awareness, and how this awareness translates into meaningful action—these are questions more difficult to answer. Still, Thoreau's plant-thinking provides possibilities for encounter that we have not yet imagined. Reading Thoreau's Journal and the late natural history projects during the pandemic years of 2020–21, I found new ways of encountering a world marked by climate change.

In the winter of 2019–20, my wife Julianne and I moved from Fairbanks, Alaska, to a new home in northern New Mexico, isolated from the coronavirus and its effects but clearly affected by the increasing heat and intensifying drought of global warming. Unlike Thoreau's long-abiding knowledge of Concord, the neighborhood of the Jemez Mountains was completely new to me, and knowledge of the place was not a birthright. In December 2019, we drove thousands of miles through snow and ice on the Alaska-Canada Highway and ultimately moved into a rose stucco house in the Jemez River Valley. Two weeks after we began living in the Jemez, the World Health Organization declared the coronavirus COVID-19 a pandemic; on 13 March

2020, the Trump administration declared the pandemic a national emergency. We found ourselves in a beautiful, remote, safe place, far from most viral dangers, but subject to the psychological and emotional effects of isolation. For recent immigrants from Interior Alaska, the temperatures of February and March were amazingly mild. As winter gave way to spring, however, we learned that my summer work as a hiking guide in Alaska had been canceled. Other summer teaching gigs were postponed until 2021, then 2022. We realized that we were going to be passing many months in the Jemez. That was where we found ourselves through the winter of 2021, edging toward and then passing an anniversary of sorts.

Moving through those seasons, I attempted to follow Thoreau's practices as I read them in the pages of the Journal. Partly this was to practice "the art of Walking" (*Exc* 185). I found myself repeating hikes and revisiting areas of the Valles Caldera National Preserve, twenty miles north of our new home. For over a year, as it turned out, I made a series of more than one hundred excursions to places that became, over the months, ever more familiar. One was the Banco Bonito, a rough plateau of rocky, volcanic ridges and meadows, covered in ponderosa pines and Douglas firs, ranging from 8,000 feet to 8,700 feet in elevation. Another was the Redondo Meadows, a large, open depression fed by Redondo Creek, running off the high ridges of Redondo Peak, at over 11,000 feet the highest mountain in the preserve.

In the Journal, the areas Thoreau names Conantum or Andromeda Ponds, Clematis Brook or Yellow Birch Swamp, are transformed by his prose into gorgeous gardens of wildflowers, swamps, and flooded meadows teeming with a huge variety of plants. The entries resound with common and scientific names of plants, becoming familiar to the writer and the reader. Shorthand, abbreviated notices abound. Skunk cabbage becomes an old friend, a first sign of spring; the willows remain a kind of ever-proliferating mystery, calling for repeated observations and descriptions. The invented place-names never feel like appropriation; instead, the naming acts as a form of appreciative recognition. And Thoreau's recognitions are always based in his observations of plants and other natural features.

Thoreau had a true genius for finding wild places near at home, and those places became even more home to him. In "Walking," he notes that "hope and the future for me are not in lawns and cultivated fields, not in towns and cities, but in the impervious and quaking swamps. . . . There are no richer parterres to my eyes than the dense beds of dwarf andromeda (*Cassandra calyculata*) which cover these tender places on the earth's surface.

Botany cannot go further than tell me the names of the shrubs which grow there—the high-blueberry, panicled andromeda,—lamb-kill, azalea—and rhodora—all standing in the quaking sphagnum" (*Exc* 204). The swamp becomes "a sacred place—a *sanctum sanctorum*," and the wild soil "is good for men and for trees" (205).

In my early excursions, I couldn't help but observe the cumulative effects of ranching, mining, logging, and wildfires on the lands of the Valles Caldera. Sacred to the ancestors and present-day members of Jemez Pueblo, many of the places I hiked had been burned in the massive Las Conchas fire of 2011 and then salvage logged. The ponderosas and Douglas firs often showed scorched trunks; dozens of dead trees stood on the blackened ground beside the trails. In the Banco Bonito, the trails formed a tangled network of old logging roads, decommissioned when the preserve was established twenty years earlier and then partly recommissioned after the 2011 fire. Much of the land felt hard used, covered in slash and logging debris, even though it still showed abundant beauty. Partly because I was a newcomer, I felt unable to name the places I visited.

The complex, western landscapes of the Valles Caldera let me imagine Thoreau looking out at the neighborhood of Concord and Lincoln, realizing that nature was now in the hands of white settlers and the beginnings of what we call "the Anthropocene." I imagined Thoreau looking at the sand foliage in the Deep Cut, formed by the railroad tracks right next to Walden Pond; Thoreau listening to the hum of the Telegraph Harp, summer and winter; Thoreau noting the logging operations in the woods around him. Thoreau could consider a local logger like Alek Therien a kind of heroic figure, but he would also know that the loggers were working in service of timber and firewood industries. In "Walking," he registered the lack of "primitive and rugged trees" in the neighborhood and considered it reason to "shudder for these comparatively degenerate days of my native village" (*Exc* 206). In the farther excursions to Maine, he saw vast forests in danger of becoming matchsticks. From my side, I could see how far our America had traveled down the industrial road in the last two centuries, how much deeper we were burning, scraping, and digging into the soil. Reading Thoreau became an exercise in surveying changes in the landscape, measuring the distance of our descent.

The plants of the Southwest I learned to identify were often beautiful and plentiful, but the number of species seemed in no way abundant. That was partly a measure of Thoreau's expertise and my lack of it, but it also seemed to point toward a reduction in diversity. I became used to seeing scraggling hosts of yarrow, yellow salsify, and dandelions. Mullein, paintbrush, penstemons, and scarlet gilia were everywhere along the trails. Over the extreme drought-sapped summer of 2020, I was astonished by how durable the *Asteraceae* proved to be, though I was prepared for more variety. Daisies, thistle, fleabanes, two species of purple aster. Rich yellows of *Rudbeckia, Senecio, Helenium, Solidago, Helianthella*. Woody shrubs sprang up all around, many of which I had never seen before. Early on, barberry and blueberry, two of Thoreau's favorites, in the high woods; by September, chamisa blooming along every road and highway.

In the Valles Caldera, as I walked the woods and meadows, I discovered that the trees were more diverse than I had initially thought. New Mexico locust saplings grabbed me constantly. Limber pine and blue spruce, Engelmann spruce and white fir, quaking aspen and Gambel's oak. I thought about Thoreau's enthusiasm for the wildflowers in May and June, but by early July extreme drought had taken hold of New Mexico, and we were praying for a decent monsoon season of thunderstorms. It never came. True, I found two stunning purple Sego lilies in the Redondo Meadows one July afternoon. In "Walking," Thoreau praises the wild in literature and claims that "a truly good book is something as natural, and as unexpectedly and unaccountably fair and perfect, as a wild flower discovered on the prairies of the west" (*Exc* 207–8). For me, the reverse was also true, and the Sego lilies proved it.

Thoreau calls, in "Walking," for a literature that truly expresses nature in all its dynamic wildness. The plants gave him a way of understanding the limits and the possibilities of language. The poet, he claims, transplants words to the page "with earth adhering to their roots," and the words would be so true that "they would appear to expand like the buds at the approach of spring, though they lay half smothered between two musty leaves in a library,—aye, to bloom and bear fruit there after their kind annually for the faithful reader, in sympathy with surrounding Nature" (*Exc* 208). Reading the Journal, I kept realizing how fresh Thoreau keeps his eye and his prose, and I practiced at being a faithful reader. But by the middle of August, I was feeling the dog days and finding myself jaded by repeated hikes, searching for new trails. Was my perception failing me? Was it my imagination? One day, I realized that one of the biggest differences between Thoreau and me might be the trail itself. Unlike Thoreau, I had been slavishly walking the dusty trails in the Caldera, rarely venturing off-trail for more than a few hundred yards. Perhaps my eyes had partly closed with every step.

In "Walking," Thoreau commands his listeners, "Give me a culture which imports much muck from the meadows, and deepens the soil, not that which trusts to heating manures, and improved implements, and modes of culture only" (*Exc* 213). With wicked irony, the technology of the Anthropocene came to my aid. I had been using a mapping download on my cellphone to guide my hikes, but now I began to use it to learn the landscape between trails, off trails, and away from trails. The map would show my location as a blue dot, blinking on the topographical lines of the ridges and peaks. The Caldera is a wilderness, but, unlike Alaska, you are never far from an old logging road or a designated trail. Under sunny skies, you could always figure out the four cardinal points of the compass.

In mid-August, I made several off-trail excursions that completely changed my experience in the Valles Caldera Preserve. It was then that the excursions began to open in new directions. On 12 August 2020, I hiked for four hours across the Banco Bonito, crossing several trails but following my own route from west to east, in that direction running counter to the strong "westward" predilection in "Walking." Along the way, I stumbled on a large bull elk, lying down on a sunny slope of ponderosas. He was up and gone in a flash of tan and dark brown, running silently away. A week later, I followed a trail above Redondo Meadows for a mile, then cut up into the rocks and climbed a thousand feet to Mirror Lake, a spring-fed pond nestled in talus at 9300 feet. The pond is beautifully well named, but the off-trail ascent and descent, following elk highways, were the highlight of the day. As they proliferated, the off-trail hikes reminded me of Thoreau's botanical excursions across lots, trespassing on land the owners had likely never walked. My excursions were becoming more like Thoreau's because of the possibility of surprise.

That first encounter with the bull elk in the Banco Bonito was a fore-runner. As summer stretched into fall, fall into winter, I seemed to keep running into herds of elk. And often the elk would take me to other dis-coveries—new plants, a sudden birdsong, a mule deer with a five by five set of antlers. Like Thoreau, I kept a record of my excursions, though in no way matching his detailed and nuanced descriptions. I kept adding to my meager store of "wild and dusky knowledge," the "tawny grammar" of plants, birds, and elk leading me into new places, on the confines of the maps and their trails. Ultimately, I came to see my hikes in the light of key sentences in "Walking":

> My desire for knowledge is intermittent; but my desire to bathe my head in atmospheres unknown to my feet is perennial and constant. The highest that we can attain to is not Knowledge, but Sympathy with Intelligence. I do not know that this higher knowledge amounts to anything more definite than a novel and grand surprise on a sudden revelation of the insufficiency of all that we called Knowledge before—a discovery that there are more things in heaven and earth than are dreamed of in our philosophy. It is the lighting up of the mist by the sun. (*Exc* 215)

The sense of discovery comes out in a marvelous story in "Walking." Thoreau admonishes us to take new views, to climb a tree rather than walk about the foot of it for one's three score years and ten. And when he climbs a tall white pine near the end of June one year, he finds "on the ends of the topmost branches only, a few minute and delicate red cone-like blossoms, the fertile flower of the white pine looking heavenward" (220). Perhaps more astounding, when he carries the topmost spire to the village, none of the inhabitants—not farmers, not lumbermen, not hunters, not woodchoppers—has seen the like before, "but they wondered as at a star dropped down!" (220).

My many excursions, and Thoreau's constant faith in discovery and surprise, also remind me strongly of his search for the *Azalea nudiflora,* or Pinxter-flower, which the hunter George Melvin showed him. The 31 May 1853 Journal account is humorous, and the story is so immediately signifi-cant to Thoreau that he likens it to allegory and myth. My favorite line is Thoreau trying to persuade Melvin, who apparently had been drinking, that he should share the location: "I was a botanist and ought to know." The episode teaches the botanist a profound lesson: "The fact that a rare and beautiful flower which we never saw—perhaps never heard of, for which therefore there was no place in our thoughts, may at length be found in our immediate neighborhood" (*PJ* 6:162). Surely a flower like the Sego lily or the sudden appearance of seventy-five elk loping all around me were teaching me a similar lesson. Not that I was an expert who "ought to know," but that just out of range of my experience whole herds and fields of creatures were living their lives and waiting to find a place in a writer's thoughts.

The allegory suggests that our knowledge is always limited, like our perceptions, and that much more lies hidden in plain view than we will ever observe and clearly see. We recognize some parts of the world, and we fail to recognize even more. For Thoreau—and for me during the pandemic—this became an article of faith: there is always a surprising encounter waiting for us, perhaps a connection we will experience on the next excursion. Then there is a larger implication. We imagine that we know the state of our world, how it holds together, how it is unraveling, and what we can or can't do about it. But what do we really know? The plants I encountered over the months were hardy, parched, productive creatures. They supported bright green hummingbirds feeding so quickly I could barely keep my eyes on them, and dozens more species I never saw at all. Thoreau could look at the mysterious life of plants and the ravages of his fellow townsmen on the woodlots around him, and he could still say, "I have great faith in a seed" (*Exc* 181). Surely, I could put some faith in the plants and animals around me.

Another lesson I gleaned from a hundred excursions was to practice what Thoreau calls "the true sauntering of the eye" (*PJ* 5:344)—that is, not to be too focused on knowledge, not to be driven by an imagined goal or purpose. If I began by learning the plants and birds, I could continue by learning the elk. No doubt they would lead me back again on another path. This idea of the sauntering of the eye, I came to feel, points toward an engagement with the world—especially the earthly world—that is neither part of Western science nor part of Western humanism. The sauntering of the eye would lead to perceptions without an intentional frame set by the traditional Western dichotomy of nature and culture. The kind of knowledge gained by the sauntering of the eye would be embedded and life sustaining, the kind of lived relationship that anthropologist Tim Ingold finds modeled in traditional hunter-gatherer societies, which he contrasts sharply with Western economies of knowledge.[2] It would be quite close to plant-thinking, as outlined by Michael Marder and practiced by Thoreau.

In my encounters with the plants, the birds, and the elk, I was hunting, but without intention. I was reading signs, but without assigning meaning or purpose. Without a fixed and focused goal, I might even have been discovering a secret of mindful sauntering. Through the Journal and a hundred excursions, Thoreau had shared an unaccountable gift, one that I had never foreseen when I started out on the path defined by the pandemic and its alienations.

The practices of walking, writing, and reading are not new, but they may help us redefine our place in a world of global climate change. This does not mean that we should make peace with the new world of the Anthropocene, that we should accommodate ourselves to it, adjust to it, and finally accept it as the status quo. Instead, the practices give us ways of knowing the new world in its details, ways of measuring our new locations and dislocations. They may give us ways to find "Sympathy with Intelligence," a surprising form of encounter. Thoreau's specific gift is the faithful, persevering record of vitality in the natural world. What we make of that gift remains open, an expectation and a surprise, but it is something in which we can responsibly place our faith.

Notes

PREFACE

1. See Sophia Thoreau to Daniel Ricketson, 20 May 1862, in Sandra Harbert Petrulionis, ed., *Thoreau in His Own Time*, 45–47.

2. The most recent work on Thoreau's encounters with plants is a collection of essays by established and emergent scholars: Branka Arsic, ed., *Dispersion: Thoreau and Vegetal Thought*. Specific essays are cited in ensuing chapters.

INTRODUCTION

1. For a stimulating overview of current research on critical plant studies, see the Literary and Cultural Plant Studies Network, which provides a list of international scholars and their current work, important research organizations and collaboratives, an excellent bibliography, and more. Much current research in the field is taking place in Canada, Europe, and Australia. The website is hosted by the University of Arizona. See https://plants.arizona.edu. A recent issue of the Australian journal *Plumwood Mountain* is devoted to "Plant Poetics," edited by John Charles Ryan (7, no. 1 [March 2020]). See https://plumwoodmountain.com/.

2. For a prescient discussion of this problem of language in Thoreau and Susan Fenimore Cooper, see Johnson, *Passions for Nature*. Johnson focuses on Thoreau's late work and his search for a language "beyond metaphor" in chapter 5 (181–218). See Johnson's recent essay on Thoreau and the New Materialism, "Materialities of Thought."

3. The Princeton Edition of Thoreau's Journal stands now at eight completed volumes, from October 1837 to 3 September 1854. This is the edition to which I refer, when possible, with the abbreviation *PJ* and volume/page number of the passage. For the Journals after September 1854, I cite the fourteen-volume edition, edited by Bradford Torrey and Francis H. Allen (1906). This edition is readily available in the two-volume Dover folio format (1962). The Princeton Edition volume is cited in Arabic numbers; the 1906 edition volume is cited as *J* with Roman numerals.

4. Buell, in *The Environmental Imagination*, details Thoreau's environmental projects in *Walden*, and botany plays a role in his discussions of phenology and the seasonal in Thoreau's writing, especially in *Walden* (115–39, 219–32, 242–51). A recent

narrative of Thoreau's lifelong search for meaning in his saunters and journal-writing, with special attention to the last decade, is Robinson, *Natural Life*.

5. Occasionally, Thoreau cites other authorities. In July 1854, for instance, he cites Alphonso Wood (1810–1881) in a discussion of common milkweed, aligning Wood correctly with Gray (*J* VI:404; see also 383, 403 for other references to Wood). Most likely Thoreau is consulting Wood's *Class-Book of Botany* (1848), intended as an elementary guide. Dupree discusses the acquaintance of Wood and Gray, 169–70. For an early copy of Wood's textbook, see https://www.biodiversitylibrary.org/page/5286596#page/1/mode/1up. In *Bird Relics*, Arsic connects Bigelow's *American Medical Botany* to Thoreau's vitalism and to a sense of the pathological in all organic life (236–43). In general, and especially in the early years of his studies, Thoreau uses the *Florula Bostoniensis* more often as a manual.

6. For the modern student, the definitive botanical textbook is Gleason and Cronquist, *Manual of Vascular Plants of Northeastern United States and Adjacent Canada*, 2nd ed. The dichotomous keys and descriptive writing, as well as the technical glossary and indexes to common and scientific names, make this an indispensable one-volume guide. When needed, I cite this text as "G&C" with page number for modern identifications. Gleason and Cronquist discuss skunk-cabbage as *Symplocarpus foetidus*, 648–49.

7. Asa Gray's career as an academic botanist is detailed in the definitive biography *Asa Gray: American Botanist, Friend of Darwin*, by A. Hunter Dupree. While many readers know Gray as the scientific adversary of Louis Agassiz and the American proponent of Darwin's *Origin of Species* and the theory of evolution by natural selection, his years as the Fisher Professor at Harvard University and his numerous textbooks are even more important in thinking about Thoreau's self-education as a practicing botanist. Gray, though an orthodox thinker and scientist, works in ways that resemble the methods of Alexander von Humboldt, with a vast network of correspondents, sharing herbarium specimens internationally and working slowly on the technical aspects of plant physiology and taxonomy. As Dupree shows in fine detail, Gray becomes a significant American ally for Darwin in the controversies surrounding the *Origin of Species*, even though Gray wished to maintain arguments for the role of a Creator and the idea of intelligent design of the material world (233–306). For a sharp view of Gray from another viewpoint, see Irmscher, *Louis Agassiz*, 121–67.

8. The Biodiversity Heritage Library is the world's largest digital collection for biodiversity literature and archives. The website is open access, and it provides a host of materials for the study of biodiversity and the history of biological sciences. For an introduction, see https://www.about.biodiversitylibrary.org. Gray's *Manual of Botany for the Northern United States* was first published in 1848 and went through multiple editions through the nineteenth and twentieth centuries. See https://www.biodiversitylibrary.org/page/10955275. For the 1850 edition of Gray's *The Botanical Text-Book*, which also went through multiple editions in the nineteenth and twentieth centuries, see https://www.biodiversitylibrary.org/page/19048186.

9. The index to the 1906 *Journal* is, of course, not exhaustive. For example, even though Thoreau cites Alphonso Wood several times, Wood does not appear in the

index at all. The number of entries for plants, moreover, certainly includes some repetitions and overlapping terms, and my survey is subject to my own errors. As for the Harvard University Herbarium, the 1295 entries include a significant element of redundancy in the species collected for an academic herbarium. Despite these disclaimers, the numbers clearly indicate Thoreau's seriousness as a student of botany.

On the front flyleaf of Thoreau's copy of Gray, *Manual of the Botany of the Northern United States*, 2nd edition (1856), Thoreau wrote, "Sep. 11–61 I have collected a little over 900 (910) flowers—not counting sedges grasses &c." (William Howarth, *The Literary Manuscripts of Henry David Thoreau* 322). The one-sheet manuscript (F28e in Howarth's notation) is in the Berg Collection in New York Public Library, laid into Gray's *Manual* and described by Howarth as Thoreau's "final checklist of flowers, compiled from previous listings." The five manuscripts for [Nature Notes: Flowers] run 200 pages (Howarth 320–22).

Walls, in *Seeing New Worlds*, quotes Walter G. Harding's count of "more than eight hundred of the twelve hundred known species of Middlesex County" and the herbarium of "more than one thousand pressed plants" (136; see Harding, *Days of Henry Thoreau* 290, 265).

The best current account of Thoreau's identifications and collections for his herbarium is Angelo, "Botanical Index to Thoreau's Journal," http://www.ray-a.com /ThoreauBotIdx/index.html. Angelo includes a clear introduction to the index, links to several of his essays, including "Thoreau as Botanist," an excellent account of Thoreau's development and the colleagues and friends who contributed to it. Angelo asserts that Thoreau collected some 900 species of plants and collected at least 1500 specimens for his herbarium. See also the introduction to the link "Herbarium Spreadsheet" on Angelo's "Botanical Index" website.

10. The number of pages refers to the 1906 edition of the Journal. In the Princeton Edition the numbers are similar, and they reflect the same waxing and waning of volume in the Journal entries.

11. See Walls, *Seeing New Worlds*, 84–93, for Alexander von Humboldt's six principles of empirical holism; 140–47, for Thoreau's version of the six principles. Walls's study of Thoreau in relation to nineteenth-century natural science, especially in relation to Humboldt and Darwin, is indispensable. Walls shows in detail how Thoreau moves to an "epistemology of contact" during the years 1850–51 (116–30; 147–57). My argument focuses more narrowly on the role of botany in Thoreau's development as a writer, but *Seeing New Worlds* is an essential book for understanding even that narrow topic. For a larger view of Humboldt in relation to American art and culture, see Walls, *Passage to Cosmos;* Thoreau makes a significant appearance in relation to other writers and artists of nineteenth-century America, 251–301. A recent account of Thoreau in relation to the physical sciences such as geology is Thorson, *Walden's Shore*.

12. For a fascinating and very different interpretive approach to Thoreau's phenology, see Case, "Beyond Temporal Borders." Case connects the Kalendar charts of "general phenomena" and "all phenomena" to Thoreau's processing of grief and the role of music. Her reading of the Kalendar charts connects them to the entries of

the Journal, as does mine, but she goes much farther in seeking to read the Kalendar charts as significant in their own right—in "Beyond Temporal Borders," as a kind of musical score.

13. See the link to "Thoreau Place Names" on Ray Angelo's website: http://www.ray-a.com/ThoreauBotIdx/index.html. A recent article in the *Thoreau Society Bulletin*, no. 314 (Summer 2021) updates Angelo's work on the place names: ray-a.com/ThoreauPlaceNames.pdf. Angelo lists place names, annotates their locations, and gives useful information about Journal entries in which they appear.

1. THE TWO BOTANICAL EXCURSIONS OF *THE MAINE WOODS*

1. Fedorko offers persuasive evidence that Sophia was alone responsible for editing Thoreau's posthumous volumes. See "Henry's Brilliant Sister," 222–56. See also *Excursions*, "Historical Introduction," 330–63. For a wide-ranging essay on Thoreau's excursions, see Schulz, "Nature, Knowledge," 30–39.

2. Thoreau lectured on "Ktaadn" in January 1848 and published his account in five installments in Sartain's *Union Magazine of Literature and Art* (July–November 1848). He lectured on "Chesuncook" in December 1853 and published the narrative in three issues of the new *Atlantic Monthly* (June–August 1858). He lectured at least once on "The Allegash and East Branch," in January 1858. The third narrative and the appendix remained unpublished until 1864. For excellent accounts of the three excursions and the essays that resulted, see Walls, *Thoreau: A Life*, 223–28, 246; 334–41; 406–23; 496–97. Moldenhauer gives a good history of the text in his textual introduction to the Princeton edition of *Maine Woods*, 355–77. Moldenhauer's excellent essay, "Maine Woods," offers the benefits of his expertise in accessible form.

3. For a critical account of Thoreau's "Indian Books" and his view of Native Americans, see Bellin, "In the Company of Savagists." Bellin's reading of *The Maine Woods* (22–26) is especially strong in its critique of earlier scholars' accounts. For a list of significant earlier accounts, see n. 15. A recent, excellent treatment of the Indian Notebooks in relation to *The Maine Woods* and "Walking" is Kucich, "Imperfect Indian Wisdom," 9–14.

4. For a graphic image of Humboldt's idea, see Walls, *Passage to Cosmos*, 44. The zonal chart reappears on the cover of Humboldt's *Essay on the Geography of Plants* (1807), edited by Stephen T. Jackson. See also Humboldt, "Ideas for a Physiognomy of Plants" in *Views of Nature*, ed. Stephen T. Jackson and Laura Dassow Walls, 155–241.

5. Kimmerer, *Braiding Sweetgrass*, 214. The following quotations are taken from the chapter "In the Footsteps of Nanabozho: Becoming Indigenous to Place" (205–15).

6. Willis appears in *Collections of the Maine Historical Society* 4 (1856), with supplementary material by C. E. Potter. See *Maine Woods*, ed. Jeffrey S. Cramer, 299–304 and bibliography. Joseph Attean, Thoreau's twenty-four-year-old Penobscot guide on the "Chesuncook" excursion of 1853, is repeatedly named as "Joe Aitteon" in the printed text; similarly, Thoreau consistently misspells the Allagash River in the 1857 excursion. It seems a good idea to use the correct spellings in writing about Thoreau.

7. The entries in the Journal support this point about Thoreau's botanical eye. Much of the description in the Journal is given to plants—both those that are encountered and those that are missed. See *PJ* 7:40–99.

8. A digital version of the manuscript is at https://iiif.lib.harvard.edu/manifests/view/drs:50012606$4i.

9. This tale of editorial hubris, often told, may be found in Walls, *Life* (422–23), where she notes that Lowell turns "Thoreau's statement of religious principle into a poet's mere personal preference" (422).

10. Even a first account of Ahimsa would embrace Hinduism, Yoga, Jainism, and Buddhism. For Emerson's and Thoreau's enthusiastic knowledge of Hindu literature and Eastern ideas, see Hodder, "Asian Influences," 27–37. The definitive study of Thoreau's spirituality is Hodder, *Thoreau's Ecstatic Witness*.

Vitalism is a philosophical concept that can be traced at least to Aristotle and to the pre-Socratic Sophists. At its most basic level of definition, vitalism holds that there is a vital force, beyond the material explanations of chemistry and physics, that accounts for life. The concept continues to spark much debate in a number of intellectual domains. For Thoreau studies, and for "materialist vitalism" as an ontological project in Thoreau, see Arsić, *Bird Relics*, especially her account of Harvard vitalism and its influences on Thoreau's thinking about a philosophy of life (117–248). Interestingly, Jacob Bigelow figures prominently as a Harvard vitalist, though Arsic focuses on Bigelow's *American Medical Botany* rather than his *Florula bostoniensis* (135–38, 194–97, 236–40). See also Rochelle Johnson's recent essays on Thoreau and the New Materialism: "This Enchantment Is No Delusion," 606–35; and "Materialities of Thought," 114–32. A good summary of transcendental materialism and Thoreau's interest in Native Americans appears in Kucich, "Imperfect Indian Wisdom," 12. For an excellent argument connecting Thoreau's materialism with racial politics, see Ellis, *Antebellum Posthuman*, 61–95.

11. The Aroostook War involved the boundary dispute between Great Britain and the United States after the Revolutionary War. The land lying on the border of northern Maine and New Brunswick was occupied by white settlers and loggers from both countries, and only in 1842 was the dispute settled. There was no actual war, but militias and civilian groups did engage in significant saber-rattling. There are comic elements in the whole story: the only deaths were two Canadian militiamen attacked by bears. The Penobscot War, on the other hand, is more ambiguous. Thoreau may be referring to the series of wars between white settlers and Native Americans in the seventeenth and eighteenth centuries, usually called the Abenaki wars, or he may be referring more pointedly to the so-called Penobscot Expedition of 1779, in which the American fleet was destroyed by the British in Penobscot Bay and Penobscot River. Cramer, in the Annotated *Maine Woods*, notes that Thoreau could be referring to the fourth Abenaki war, "part of Dummer's War (1721–26), which included Lovewell's Fight and the death of Father Sebastian Rasles" (118n134). For an account of several of the wars between Northeast Natives and white settlers, as well as the efforts of allied Native peoples to negotiate treaties with the colonists and the United States, see Brooks, *Common Pot*, 51–218.

12. Joseph Attean was twenty-four years old at the time and, according to Thoreau, appeared to identify himself with lumbermen more than with traditional Penobscot lifeways (*MW* 90, 107). A recent film features more biographical detail on Attean, the son of governor John Attean and hereditary governor himself for seven years up until to his drowning death on July 4, 1870: "The Penobscot," https://vimeo.com/427895797. See also Fannie Hardy Eckstorm, *The Penobscot Man* (1904). In *The Common Pot*, Brooks portrays Attean as already a leader in his time with Thoreau (247).

13. "The Nocturnal Wildlife of the Primeval Forest," *Views of Nature*, 140–48. Thoreau uses the word "primitive," while the modern translation of Humboldt employs "primeval." In context, these terms appear to be synonyms.

14. See Cronon, *Changes in the Land*, 54–81.

15. Cramer identifies the allusion to Dickens's "Noble Savage" in *Household Words* (*Annotated Maine Woods* 145). Cramer's scholarship in this volume is invaluable.

16. Joseph Polis has long fascinated readers of Thoreau. Significant scholarly interpretations include Gura, "Thoreau's Maine Woods Indians"; Kucich, "Lost in the Maine Woods"; Lynch, "Domestic Air' of Wilderness"; Papa, "Reinterpreting Myths"; and Sayre, *Thoreau and the American Indians*.

17. In *The Life and Traditions of the Red Man* (1893), Joseph Nicolar recounts a similar traditional story of Klose-kur-beh's hunting and killing the first moose. See the modern edition, ed. Kolodny, 130–33. Kolodny's scholarly "History of the Penobscot Nation" and "Introduction" give many valuable insights into Nicolar's book and its place in Native American literature.

18. Cramer notes the allusion to John 1:5: "And the light shineth in the darkness" (*Annotated Maine Woods* 168).

19. Kimmerer, "Weaving Traditional Ecological Knowledge," 434. For the entire article and bibliography, see 432–38.

20. Brooks, *Common Pot*, 8–9. Brooks's discussion of the writing networks of Wabanaki peoples is wide ranging and resonates in many ways with Thoreau's desire to retrace the waterways of the Maine woods. The entirety of *The Common Pot* sheds light on Thoreau's insights and limitations as a traveler among the Penobscot. For example, she notes Thoreau's privileging of the oral in his night with the Penobscot and St. Francis Indians in "Chesuncook" (247). Thoreau uses an approximate spelling of the term for writing in that same episode, "*wighiggin*" (*MW* 137) and translates the term as "a bill," a kind of passport. The "Chesuncook" episode is another of Thoreau's near misses, in which he grazes the relationship between writing and speech.

21. See Cramer's Annotated *Maine Woods*, 185n93.

22. For Thoreau's rendering of *Prometheus Bound*, see *Translations*, 3–53. Later in the passage at hand, Thoreau revises his own translation, "rough tooth of the sea," which he had translated as "rough jaw of the sea" in 1843. This phrase occurs in Prometheus's prophecy of Io's journey to Egypt (*Translations* 36). See also Cramer, Annotated *Maine Woods* 218–20.

23. See https://iiif.lib.harvard.edu/manifests/view/drs:50012606$3i.

24. For a telling account of Polis as inhabiting the same world as Thoreau, rather than being figured as a mythic representative of wildness, see Traub, "First-Rate Fellows." This essay is especially valuable for giving the transcription of Huntington Library Manuscript HM 13199. Traub reads one manuscript passage as seeing Thoreau's Indian guides as "first-rate fellows," but the parenthetical remark seems to me to attribute that term to the "batteaumen and loggers." The passage calls the Indian guides "far more instructive companions for us than any white man could have been" (88). My reading of Polis emphasizes the idea in that last phrase.

25. Rasle was a Jesuit missionary to the Abenaki, founding a mission on the Kennebec River at the village of Norridgewock. He began work on the dictionary in 1691 and left it unfinished at his death in 1728. The manuscript dictionary was edited in 1833 by John Pickering in the *Memoirs of the American Academy of Arts and Sciences* (I, 375–574) under the title "A Dictionary of the Abnaki Language in North America by Father Sebastian Rasles." See the entry for Father Rasle in the *Dictionary of Canadian Biography.* http://www.biographi.ca/en/bio/rale_sebastien_2E.html.

2. *CAPE COD* AND THE SEVEN EXCURSIONS

1. Walls, *Life*, gives clear, succinct accounts of the first excursion, the lectures, and the essay publications in *Putnam's* (275–82, 375–77). For a more elaborate and detailed narrative, see the "Historical Introduction" by Moldenhauer in *CC* 249–96.

2. In *Thoreau beyond Borders*, ed. Specq, Walls, and Nègre, several critics focus on the themes of shipwrecks, history, and political economy. See Vogelius, "*Cape Cod*'s Transnational Bodies"; Weisburg, "Beyond the Borders of Time"; and Pickford, "*Cape Cod*, Literature" (179–94). For an essay that explicates four textual dimensions for reading Thoreau's excursions, at the same time that it offers dynamic close readings of "Walking" and *Cape Cod*, see Walls, "Walking West, Gazing East."

3. See Moldenhauer's excellent discussion in *CC* 249–77.

4. See *G & C*, 155–56 and 780–81. By the time Thoreau makes his last excursion to Cape Cod in July 1857, he seems to be confident that poverty grass is the common name for the two species of *Hudsonia*.

5. *G & C*, 594–95. A useful website for quick reference is Go Botany of the Native Plant Trust, with a focus on New England plants: https://gobotany.nativeplanttrust.org/.

6. Among these pages from the 1850 Cape Cod excursion, one finds a pair of paragraphs devoted to Native Americans. Thoreau paints a picture of remnants, "an eccentric farmer descended from an Indian Chief" or "a solitary pure blooded Indian looking as wild as ever among the pines—one of the last of the Massachusett's tribes stepping into a railroad car with his gun & pappoose [*sic*]." He then focuses on "an Indian squaw with her dog," living alone and "weaving the shroud of her race—performing the last services for her departed race. Not yet absorbed into the elements again— A daughter of the soil—one of the nobility of the land—the white man an imported weed burdock & mullein which displace the ground nut" (*PJ* 3:93). The

wavering viewpoint comes nearest to sharp focus in the final sentence, another example of plant-thinking in its early stages.

7. The quotation from Michaux appears in *The North American Sylva* 1:99. Available on the Biodiversity Heritage Library website: https://www.biodiversitylibrary .org/item/215812#page/195/mode/1up.

8. Benjamin Marston Watson (1820–1896) was a native of Plymouth, a Harvard graduate, and a fellow traveler of the Transcendentalists. Mary Russell Watson was best friends with Lidian Emerson, both women being natives of Plymouth. The Watsons were well known for their Gothic home Hillside and the eighty-acre garden and orchard adjoining it. Marston Watson was one of the founders of the Natural History Society at Harvard, and the Watsons invited Thoreau to Plymouth several times over the years to lecture on *Walden*, "Walking," and "Cape Cod." Walls, *A Life*, provides the most information about the Watsons and their friendship with Thoreau (68, 121,138, and *passim*). On one trip to Plymouth in October 1854, Thoreau gave lectures to a small group and spent three days surveying the Watsons' property (364). Walls also summarizes the June 1857 trip to Plymouth and Cape Cod (404–5).

3. *WALDEN* AS BOTANICAL EXCURSION

1. Shanley, *Making of Walden*; Clapper, "Development of *Walden*"; Sattelmeyer, *Writing the American Classics*, 53–78; Adams and Ross, *Revising Mythologies*, 162–92; Walls, *Life*, 331–55.

2. The standard scholarly edition of *Walden* is edited by J. Lyndon Shanley. I cite from that edition. Another very useful edition is *Walden, Civil Disobedience, and Other Writings*, 3rd ed., ed. William Rossi. This current Norton Critical Edition includes many helpful essays and a recent bibliography. Sattelmeyer's essay on the "Remaking of *Walden*" is reprinted there (489–507).

3. As with other instances, Ray Angelo's *Botanical Index to Thoreau's Journal* is invaluable. He identifies the "white grape" as "*Vitis labrusca* forma *alba* (FOX GRAPE)." He identifies "spoonhunt" as *Kalmia latifolia* (Mountain Laurel, or "spoon-wood" in Gray, *Manual* 319). He identifies "hoopwood tree" ("hopwood" in *PJ*) as *Fraxinus nigra*, the black ash (*Fraxinus sambucifolia* in Gray, *Manual* 336). The red huckleberry is native to the northwest coast of North America. Thoreau mentions it in a Journal entry, 2 August 1853: "John Le Grosse brought me a quantity of red huckleberries yesterday. The less ripe are whitish. I suspect that these are the *white* huckleberries" (*PJ* 6:284). Angelo interprets "red huckleberry" as an alternative for "black huckleberry," and Gray calls *Gaylussacia resinosa*, or black huckleberry, the "common *Huckleberry* of the markets" (311). Gray's *Manual* often gives common names and scientific names that may have fallen out of usage in our times, so it is an excellent source for the identifications in Thoreau's work.

4. The relevant source for the sand cherry in "Sounds" is a 12 May 1850 Journal entry: "The sand cherry (Prunus depressa. Pursh. Cerasus pumila. Mx.) grew about my door and near the end of May enlivened my yard with its umbels arranged

cylindrically about its short branches. In the fall weighed down with the weight of its large and handsome cherries it fell over in wreath-like rays on every side. I tasted them out of compliment to nature, but I never learned to love them" (*PJ* 3:68). The sentence in "Sounds" reads as follows: "Near the end of May, the sand-cherry, *Cerasus pumila,* adorned the sides of the path with its delicate flowers arranged in umbels cylindrically about its short stems, which last, in the fall, weighed down with good sized and handsome cherries, fell over in wreaths like rays on every side. I tasted them out of compliment to Nature, though they were scarcely palatable" (*W* 113–14). Bigelow lists both scientific names for the species and therefore seems to be Thoreau's source for the scientific names in the 1850 Journal passage (*FB* 193).

5. In a Journal entry from 9 May 1841, Thoreau compares the pine tree to an Indian, "with a fantastic wildness about it even in the clearings. . . . The pitch pines are the ghosts of Philip and Massassoit" (*PJ* 1:308–9).

6. For exhaustive notes on the text, see Cramer, *Walden.* I owe the reference to the 6 July 1845 entry of the Journal to Cramer's scholarship. That is one among many debts to the editor and his deep knowledge of the work.

7. For a narrative reflection on these fundamental issues of food, economy, and ecology, as well as the story of how Thoreau prepared for the move to Walden Pond, see Walls, *Life,* 181–207.

8. *Walden* 252. Cramer notes in the Annotated *Walden* that the poem was originally published in the April 1843 issue of the *Dial* (243n68).

9. See, for example, Thorson, *Walden's Shore,* 278–88. An earlier, more statistically inflected study that makes a similar point is Adams and Ross, *Revising Mythologies,* 51–63, 165–91; see especially 175. Both studies use the Journal to show how Thoreau revised the second half of *Walden* in 1852–54. The views of two eminent Thoreau biographers are precise: see Richardson, *Life of the Mind,* on the "triumph of the organic" (310–13); and Walls, *Life,* for "the incandescent coming of 'Spring'" (344).

10. Adams and Ross, 175. I have identified at least sixteen passages in my research. Adams and Ross assert that "nearly all" the Journal passages date from winter and spring 1851–52, but that does not accord with my findings. To my knowledge, the first sand foliage passage appears in the Journal of spring 1848 and undergoes intensive later revision in a series of numbered passages (*PJ* 2:382–84, 576–78). In addition, I have found ten passages from the Journals of December 1851 to April 1852, some of which merit extensive discussion (4:230–31, 285, 293, 294, 302, 363 383, 384, 388; *PJ* 5:6). Finally, in February–March 1854, four important additions appear as Thoreau is making the final revisions of *Walden* (*PJ* 7:268–69, 276; *PJ* 8:25–26, 30). For the most recent account of the manuscripts, see William Rossi, "Making *Walden* and Its Sandbank," *Concord Saunterer,* 30 (2022): 10–58.

11. The *Oxford English Dictionary* has good quotations for "lights" meaning hog's lungs and used in combination with pieces of liver. See this colloquial expression from Mark Twain's *Adventures of Huckleberry Finn* (1884): "It most scared the livers and lights out of me." The figurative meaning of "bowels" as compassion, pity, or "heart" was current in Thoreau's day, though perhaps less so now (*OED*).

12. These terms are included in the glossary of Gray's *Manual of Botany* and in the glossary of Gleason and Cronquist's *Manual of Vascular Plants.* Thoreau studied lichens assiduously in the winter of 1851–52. On 31 December 1851, for instance, the same entry in the Journal as the "liver-lights and bowels" of sand foliage records "a good day to study lichens" and describes "an exhibition of lichens" in the forest (*PJ* 4:231–32). On 26 January 1852, Thoreau remarks on "the beauty of lichens with their scalloped leaves—the small attractive fields—the crinkled edge" (4:293). The collocation of descriptions in the Journal suggests that Thoreau's botanical interests were reaching new areas.

13. In the *Botanical Text-Book,* Gray discusses lichens in terms of Linnaean classification, along with ferns, mosses, and mushrooms, but the symbiotic relationship forming the lichen was not known to Western science until 1867, when Swiss botanist Simon Schwendener hypothesized the dual structure of fungus and alga. Thoreau summarizes his identifications of some 30 species of lichen from the winter of 1852 in the entry for 18 June 1852 (*PJ* 5:107–10).

14. For the discussion of the phonemes of language, Thoreau owes an intellectual debt to the Hungarian philologist Charles Kraitsir, who was an influential friend of Elizabeth Peabody and who wrote extensively on the internal and external "germs" of language. See Michael West, 262–74, and Gura, *Wisdom of Words* 109–44. For another account of Thoreau's theory of language and his views of eloquence, see Warren, *Culture of Eloquence* 53–84. For a sharp account of Kraitsir in political context, see Gura, *American Transcendentalism* 233–34.

4. THOREAU'S KALENDAR

1. As noted earlier, Case has done the most intellectually challenging and adventurous interpretations of the Kalendar. Note most recently "Beyond Temporal Borders." The illustrations to "Beyond Temporal Borders" include some charts for the first time in published form.

2. In *Life of the Mind,* Richardson details this process in a clear paragraph: "Taking advantage of his annual upsurge of energy in March [1860], Thoreau began a massive reordering of his journal and extract books. Over the next twenty-two months, extending through January 1862, he worked at rearranging the natural history materials and observations he had been accumulating systematically for ten years. His working procedure now was to run through his journal entries for a single month, say April, of a single year, making a list of observations in a single category, such as leafing, in chronological order. Then he would go on to April of the next year and write down all the leafing data for that month. From nine or ten such lists, generally beginning with 1852, he would then compile a large chart, enabling him to track each item of April leafing across ten Aprils. He repeated the entire procedure for flowering, again for bird sightings, again for different fruits, for quadrupeds, and for fish. Eventually he accumulated over 750 pages of these lists and charts, some of which must have taken many days to complete. It is a huge undertaking, a major

effort, the general purpose of which seems to have been the distillation of ten years' observations into an archetypal year, not impressionistic, but statistically averaged, combining the accuracy of a Darwin with the descriptive flair of a Pliny and the eye of a Ruskin" (381).

The most important study of phenology and climate change in the Concord area, based on Thoreau's Journal and Kalendar and extended through to the present, is Primack, *Walden Warming*. The Primack Lab website gives ample information about the phenological studies conducted by members of the research laboratory and the ways in which it has grown over the last decade (rprimacklab.com).

3. Case, "Knowing as Neighboring," 107–29. The digital archive of the Kalendar, featuring pages of "General phenomena" for April, May, June, October, November, and December, is in progress at Thoreau's Kalendar (https://thoreauskalendar.org /index.html). As Case and her coworkers present more digital versions of the voluminous materials we call the Kalendar, new interpretations are bound to emerge.

4. Peck, *Thoreau's Morning Work*, 45–49, 79–114.

5. See Howarth, *Literary Manuscripts*, F17–32 ("Nature Notes," 306–31).

6. See Howarth, *Literary Manuscripts*, F28a (320–21). "Nature Notes: Flowers" is a 42-side compilation of notes on flowers from the Journal.

7. George B. Emerson, *A Report on the Trees and Shrubs*, https://www .biodiversitylibrary.org/page/42455890.

8. Howarth describes two manuscripts that are pertinent: F18a and F18b both give material on March. The latter, at the Huntington Library, is titled "Calendar for March" and features nine pages of phenomena taken from the Journal, including the note that sleighing ends on the first of the month.

9. The key word is of course "sauntering," which Thoreau discusses at the beginning and end of "Walking." The history of this seminal essay and its readers goes beyond the confines of the present discussion. See the essay in *Exc* 185–222. For a good account of the essay as a lecture and as Thoreau's first posthumous publication in the *Atlantic Monthly*, see Moldenhauer, "Headnote," in *Exc*, 561–70. I return to "Walking" in the epilogue.

10. Richardson gives a brief, telling account of the Kansas bloodshed and this journal passage in *Life of the Mind* (344–45). During the summer of 1856, Thoreau was preoccupied with Kansas after senator Charles Sumner's speech and caning on the floor of the Senate (19–20 May) and John Brown's violence at Pottawatomie Creek (24 May). See also Walls, *Life*, 445–56; and Ellis, *Antebellum Posthuman*, 61–95.

11. See Ray Angelo's several identifications of Andromeda species. The closest modern equivalent to the Dwarf Andromeda is *Chamaedaphne calyculata* (Leather-leaf). http://www.ray-a.com/ThoreauBotIdx/BI-A.htm. Bigelow, *Florula bostonien-sis*, 176–77. In Gray, Leatherleaf is classed as *Cassandra calyculata* (317–18).

12. *The Daily Henry David Thoreau: A Year of Quotes from the Man Who Lived in Season*. A deep reflection on Thoreau's turn toward seasonal spirituality, including a subtle reading of "Ktaadn," is Robinson, "Thoreau: Crossing to the Sacred," in *Thoreau beyond Borders*, 213–25.

13. See the "Headnote" to "Autumnal Tints" in *Exc* 601–9.

14. Moldenhauer counts 30 extracts from fall 1857, 80 from fall 1858. The total is 175 (*Exc* 601).

5. *THE DISPERSION OF SEEDS* AND THE WRITER'S FAITHFUL RECORD

1. The headnote to the "Address" in *Excursions* includes these details (544–45).

2. Interestingly, Bradley Dean's publication of *Faith in a Seed* pre-dates his dissertation, "A Textual Study of Thoreau's 'Dispersion of Seeds' Manuscript." In the introduction, he notes that the published edition of the manuscript is 150 pages long and is comprised of three chapters. The first chapter is 84 pages long and treats the means by which seeds are dispersed (wind, water, and animals). The second chapter is 48 pages long and treats the succession of forest trees. The third chapter is 23 pages long and treats forest management. (Dean, "Textual Study," 4–6). *Faith in a Seed* is a judiciously edited version of the manuscript and is my proof text for *The Dispersion of Seeds*. For an excellent description of the "Dispersion of Seeds" manuscript, see Howarth F30.

3. Walls, *Seeing New Worlds*, 183–211. In addition to the Darwinian influence, Walls shows that "Succession of Forest Trees" is a canny performance that employs a number of rhetorical techniques to disrupt the illusion of scientific objectivity and create a form of relational knowing, 199–211. Walls revisits the address and *Dispersion of Seeds* in the biography of Thoreau (*A Life* 457–75). Three very recent essays on *The Dispersion of Seeds* appear in Arsic, *Dispersion*: Allewaert, "Green Fire" (105–26); Noble, "Riddle" (127–41); and Gladstone, "Low-Tech Thoreau" (143–64).

4. Darwin, *On the Origin of Species*, 425.

5. Richardson gives a succinct account in his introduction to *Faith in a Seed* (3–17, esp. 11–16). In *Life*, Walls details the initial "Darwin dinner" and Thoreau's careful study of the single copy of *Origin of Species* in January and February 1860 (457–61).

6. See Headnote to "An Address on the Succession of Forest Trees" (*Exc* 544–55); for a concise account of the controversy with Greeley, see Walls, 472.

7. Walls, 274–75. The Journal entry is from 18 October 1841 (*PJ* 1:191). See also Walls, *Seeing New Worlds*, 42–44.

8. Quammen, *Song of the Dodo*, gives a perceptive account of Darwin and the development of biogeography as a science. See also the first volume of the definitive biography by Browne, *Charles Darwin: Voyaging*, 163–340.

6. *WILD FRUITS* AND TRANSFORMATIVE PERCEPTIONS

1. Dean's account of the manuscript reconstruction appears in the note to the title *Wild Fruits* (WF 287). See also Howarth, B196; F29b; F29e; F29h; F29j.

2. In "Language of Prophecy,'" Fink interprets "Wild Apples" persuasively as a familiar lecture / essay, a genial, accessible treatment of a topic for a popular audience, but one that also probes deeply into philosophical, social, and literary topics at the

same time. This reading of "Wild Apples" is also valuable for giving the sense of Thoreau's late work as a hybrid. Newman argues that *Wild Fruits* evokes a materialist, communal vision in contrast to the transcendentalist, individualist view of *Walden*, see his book *Our Common Dwelling*, especially 171–83; see also his synoptic essay "Thoreau's Materialism," 105–37. For the larger biographical context of "Wild Apples" and *Wild Fruits*, see Walls, *Life*, 457–500.

3. See Dean's discussion of the relationship in his note to the *Wild Fruits* introduction (*WF* 287–88). He had originally identified another beginning for the *Wild Fruits* text, but he decided later that this was one of two possible beginnings for the *Dispersion of Seeds* text. Since Thoreau was heavily involved in both projects in October–November 1860, the Journal passages often elide the distinction between the two projects and texts.

4. See especially Newman, "Thoreau's Materialism," 116–30. Newman rightly emphasizes the role of the huckleberry parties and Thoreau's active leadership of them as a practical example of community-building. That makes the introductory image of "some child's first excursions a-huckleberrying" (*WF* 4) all the more significant.

5. The passage also clearly echoes passages in *Walden* and "Life without Principle." In "The Village," Thoreau has a long paragraph detailing the virtues of getting lost: it helps us "appreciate the vastness and strangeness of Nature," and "not till we have lost the world, do we begin to find ourselves, and realize where we are and the infinite extent of our relations" (*W* 171). "Life without Principle" was titled "What Shall It Profit?" when Thoreau delivered it as a lecture in Concord on 14 February 1855 (*RP* 369).

6. Fink discusses the end of "Wild Apples" as a biblical jeremiad, intended both for a hasty reader and a deliberate one, in "Language of Prophecy" (226–30).

7. Thoreau's social and political reform writings are gathered in *Reform Papers*. "Life without Principle" (*RP* 155–79) is especially pertinent here because it was one of the last four lectures that Thoreau prepared for publication in the *Atlantic Monthly* at the end of his life. The demanding tone of the essay may strike some readers as rebarbative, but it was Thoreau's most frequently delivered lecture (*RP* 369). As a direct statement of his position against the prevailing business-minded, newspaper-reading public and the immoral political institutions supported by these mental habits, "Life without Principle" is a useful gloss on the more indirect, even jovial tones Thoreau adopts in essays like "Wild Apples" or addresses like "Succession of Forest Trees."

8. Dean discusses Thoreau's plans to join this entry with the "Late Whortleberry" and "Hairy Huckleberry" entries to form a single lecture/essay like "Wild Apples," but the writer did not live to complete his revisions. For these reasons, Dean restores the "Black Huckleberry" entry as well as the other two. See Dean's note (*WF* 295–96). The Journal provides ample evidence that Thoreau was drafting a lecture on "The Huckleberry" from November 1860 through January 1861 (*J* XIV:273–310). On 11 January 1861, he writes, "I presume that every one of my audience knows what a huckleberry is,—has seen a huckleberry, gathered a huckleberry, and, finally, has tasted a huckleberry,—and that being the case, I think that I need offer no apology if I make huckleberries my theme this evening" (310). Howarth lists two manuscripts

on the huckleberry, notes compiled from the *Journal*, 1850–58 (F29a and F29b). The entry for "Black Huckleberry" appears in the "Notes on Fruits" manuscript (Howarth F29h), leaves 526–48.

9. "Whortleberry" is chiefly British, referring to *Vaccinium myrtillus* and, by extension, to the whole genus *Vaccinium*. The earliest usage dates to the sixteenth century (*OED*). Thoreau discusses the whortleberries and the names given to them in an early paragraph of "Black Huckleberry" (*WF* 41).

10. See Dean's note (*WF* 322) and the 20 November 1853 Journal entry (*PJ* 7:167–70). The difficulties and financial losses Thoreau encountered in publishing *A Week on the Concord and Merrimack* Rivers are detailed in Walls, *Life,* 254–72.

11. "[Journal extracts] Draft of 'Huckleberries." The material in this manuscript appears in altered order in Leo Stoller's edition of "Huckleberries," as well as in subsequent printings of the lecture / essay.

12. Emerson delivered the eulogy at Thoreau's funeral and published it in the August 1862 issue of the *Atlantic Monthly* (239–49). See also Myerson, "Emerson's 'Thoreau,'" 17–92.

EPILOGUE

1. Primack, *Walden Warming.*
2. *The Perception of the Environment* 40–60.

Bibliography

Adams, Stephen, and Donald Ross. *Revising Mythologies: The Composition of Thoreau's Major Works.* Charlottesville: University Press of Virginia, 1988.

Allewaert, Monique. "Green Fire: Thoreau's Forest Figuration." In Arsic, *Dispersion,* 105–26.

Angelo, Ray. "Botanical Index to Thoreau's Journal." http://www.ray-a.com /ThoreauBotIdx/index.html.

Arsić, Branka. *Bird Relics: Grief and Vitalism in Thoreau.* Cambridge, MA: Harvard University Press, 2016.

Arsić, Branka, ed. *Dispersion: Thoreau and Vegetal Thought.* New York: Bloomsbury, 2021.

Bellin, Joshua David. "In the Company of Savagists: Thoreau's Indian Books and Antebellum Ethnology." *Concord Saunterer* 16 (2008): 1–32.

Berger, Michael. "Henry David Thoreau's Science in *The Dispersion of Seeds.*" *Annals of Science* 53 (1996): 381–97.

———. *Thoreau's Late Career and* The Dispersion of Seeds: *The Saunterer's Synoptic Vision.* New York: Camden House, 2000.

———. "Thoreau's Third Book: Literature, Science, and Epistemology in *The Dispersion of Seeds* and Other Writings of the Late Career." PhD diss., University of Cincinnati, 1997.

Bigelow, Jacob. *American Medical Botany: Being a Collection of the Native Medicinal Plants of the United States, Containing Their Botanical History and Chemical Analysis, and Properties and Uses in Medicine, Diet, and the Arts.* 3 vols. Boston: Cummings & Hilliard, 1817–20.

———. *Florula Bostoniensis: A Collection of Plants of Boston and Its Vicinity, with Their Generic and Specific Characters, Principal Synonyms, Descriptions, Places of Growth, and Time of Flowering, and Occasional Remarks.* 2nd ed. Boston: Cummings & Hilliard, 1824.

———. *Florula Bostoniensis.* 3rd ed. Boston: Little & Brown, 1840.

Brooks, Lisa. *The Common Pot: The Recovery of Native Space in the Northeast.* Minneapolis: University of Minnesota Press, 2008.

Browne, Janet. *Charles Darwin: Voyaging.* Princeton, NJ: Princeton University Press, 1995.

Buell, Lawrence. *The Environmental Imagination: Thoreau, Nature Writing, and the For-mation of American Culture.* Cambridge, MA: Harvard University Press, 1995.

Cameron, Sharon. *Writing Nature: Henry Thoreau's Journal.* New York: Oxford University Press, 1985.

Case, Kristen. "Beyond Temporal Borders: The Music of Thoreau's Kalendar." In Specq, Walls, and Nègre, *Thoreau beyond Borders,* 147–62.

———. "Knowing as Neighboring: Approaching Thoreau's Kalendar." *J19: The Journal of Nineteenth-Century Americanists* 2, no. 1 (Spring 2014): 107–29.

———. "Phenology." In Finley, *Context,* 259–68.

———. "Thoreau's Radical Empiricism: The Kalendar, Pragmatism, and Science." In Specq, Walls, and Granger, *Thoreauvian Modernities,* 140–49.

Clapper, Ronald Earl. "The Development of *Walden*: A Genetic Text." PhD diss., University of California, Los Angeles, 1967.

Cramer, Jeffrey S., ed. *The Maine Woods: A Fully Annotated Edition.* New Haven, CT: Yale University Press, 2009.

———. *Walden: A Fully Annotated Edition.* New Haven, CT: Yale University Press, 2004.

Cronon, William. *Changes in the Land: Indians, Colonists, and the Ecology of New England.* New York: Hill & Wang, 1983.

Darwin, Charles. *Journal of Researches into the Natural History and Geology of the Countries Visited during the Voyage of H.M.S. Beagle round the World: Under the Command of Capt. Fitz Roy, R.N.* New York: Harper & Brothers, 1846.

———. *On the Origin of Species by Means of Natural Selection.* New York: D. Appleton, 1860.

Dean, Bradley P. "A Textual Study of Thoreau's 'Dispersion of Seeds' Manuscript." PhD diss., University of Connecticut 1993.

Dean, Bradley P., ed. *Faith in a Seed: The Dispersion of Seeds, and Other Late Natural History Writings.* Washington, DC: Island, 1993.

———. *Wild Fruits: Thoreau's Rediscovered Last Manuscript.* New York: W. W. Norton, 2000.

Dupree, A. Hunter. *Asa Gray: American Botanist, Friend of Darwin.* Baltimore, MD: Johns Hopkins University Press, 1959.

Eckstorm, Fannie Hardy. *The Penobscot Man.* Boston: Houghton, Mifflin, 1904.

Ellis, Cristin. *Antebellum Posthuman: Race and Materiality in the Mid-Nineteenth Century.* New York: Fordham University Press, 2018.

Emerson, George B. *A Report on the Trees and Shrubs Growing Naturally in the Forests of Massachusetts.* Boston: Dutton & Wentworth, 1846.

Emerson, Ralph Waldo. *Nature.* In *Essays and Lectures.* New York: Library of America, 1983.

———. "Thoreau." *Atlantic Monthly,* August 1862, 239–49.

Fedorko, Kathy. "'Henry's Brilliant Sister': The Pivotal Role of Sophia Thoreau in Her Brother's Posthumous Publications." *New England Quarterly* 89, no. 2 (June 2016): 222–56.

Fink, Stephen. "The Language of Prophecy: Thoreau's 'Wild Apples.'" *New England Quarterly* 59, no. 2 (1986): 212–30.

Finley, James S., ed. *Henry David Thoreau in Context.* Cambridge: Cambridge University Press, 2017.

Gladstone, Jason. "Low-Tech Thoreau; or, Remediations of the Human in *The Dispersion of Seeds.*" In Arsic, *Dispersion* 143–64.

Gleason, Henry A., and Arthur Cronquist, *Manual of Vascular Plants of Northeastern United States and Adjacent Canada.* 2nd ed. New York: New York Botanical Garden, 1991.

Gray, Asa. *The Botanical Text-book, an Introduction to Scientific Botany, Both Structural and Systematic. For Colleges, Schools, and Private Students.* New York: Putnam, 1850.

———. *Manual of Botany for the Northern United States, from New England to Wisconsin and South to Ohio and Pennsylvania Inclusive (the Mosses and Liverworts by Wm. S. Sullivant,) Arranged According to the Natural System.* Boston: J. Munroe, 1848.

———. *Manual of Botany for the Northern United States, from New England to Wisconsin and South to Ohio and Pennsylvania Inclusive (the Mosses and Liverworts by Wm. S. Sullivant,) Arranged According to the Natural System.* 2nd ed. New York: Putnam, 1856.

Gura, Philip F. *American Transcendentalism: A History.* New York: Hill & Wang, 2007.

———. "Thoreau's Maine Woods Indians: More Representative Men." *American Literature* 49 (1977): 366–84.

———. *The Wisdom of Words: Language, Theology, and Literature in the New England Renaissance.* Middletown, CT: Wesleyan University Press, 1981.

Harding, Walter G. *Days of Henry Thoreau: A Biography.* New York: Knopf, 1965.

Haskell, David George. *The Songs of Trees: Stories from Nature's Great Connectors.* New York: Viking, 2017.

Higgins, Richard. *Thoreau and the Language of Trees.* Berkeley: University of California Press, 2017.

Hodder, Alan. *Thoreau's Ecstatic Witness.* New Haven, CT: Yale University Press, 2001.

Howarth, William. *The Literary Manuscripts of Henry David Thoreau.* Columbus: Ohio State University Press, 1973.

Humboldt, Alexander von. *Essay on the Geography of Plants.* Edited by Stephen T. Jackson. Chicago: University of Chicago Press, [1807] 2010.

———. *Views of Nature.* Edited by Stephen T. Jackson and Laura Dassow Walls. Chicago: University of Chicago Press, 2014.

Ingold, Tim. *The Perception of the Environment: Essays on Livelihood, Dwelling, and Skill.* New York: Routledge, 2000.

Irmscher, Christoph. *Louis Agassiz: Creator of American Science.* New York: Houghton Mifflin, 2013.

Johnson, Rochelle L. "Materialities of Thought: Botanical Geography and the Curation of Resilience in Susan Fenimore Cooper and Henry David Thoreau." In Specq, Walls, and Nègre, *Thoreau beyond Borders,* 114–32.

—. *Passions for Nature: Nineteenth-Century America's Aesthetics of Alienation*. Athens: University of Georgia Press, 2009.

—. "'This Enchantment Is No Delusion': Henry David Thoreau, the New Materialisms, and Ineffable Materiality." *ISLE: Interdisciplinary Studies in Literature and Environment* 21, no. 3 (Summer 2014): 606–35.

Jonik, Michael. "'Wild Thinking' and Vegetal Intelligence in Thoreau's Later Writings." In Arsic, *Dispersion*, 85–103.

"[Journal extracts] Draft of 'Huckleberries,' based on Journal Entries." 1862. New York Public Library Digital Collections, Henry W. and Albert A. Berg Collection of English and American Literature. https://digitalcollections.nypl.org/items/b526a600-5623-0132-b325-58d385a7bbd0.

Kimmerer, Robin Wall. *Braiding Sweetgrass: Indigenous Wisdom, Scientific Knowledge, and the Teachings of Plants*. Minneapolis, MN: Milkweed, 2013.

—. "Weaving Traditional Ecological Knowledge into Biological Education: A Call to Action." *BioScience* 52, no. 5 (May 2002): 432–38.

Kucich, John J. "An Imperfect Indian Wisdom: Thoreau, Ecocultural Contact, and Spirit of Place." In Specq, Walls, and Nègre, *Thoreau beyond Borders*, 9–14.

—. "Lost in the Maine Woods: Henry David Thoreau, Joseph Nicolar, and the Penobscot World." *Concord Saunterer* 19/20 (2011–12): 22–52.

—. "Native America." In Finley, *Context*, 196–204.

Leach, Hadley C. "Natural History." In Finley, *Context*, 227–35.

Lynch, Thomas P. "The 'Domestic Air' of Wilderness: Henry Thoreau and Joe Polis in the Maine Woods." *Weber Studies* 14, no. 3 (1997): 38–48.

Marder, Michael. "Auto-Heteronomy: Thoreau's Circuitous Return to Vegetal Life." In Arsic, *Dispersion*, 59–68.

—. *Plant-Thinking: A Philosophy of Vegetal Life*. New York: Columbia University Press, 2013.

Michaux, François André. *The North American Sylva; or A Description of the Forest Trees of the United States, Canada, and Nova Scotia: Considered Particularly in Respect to Their Use in the Arts, and Their Introduction into Commerce; to Which Is Added a Description of the Most Useful of the European Forest Trees*. Philadelphia: R. P. Smith, 1853.

Myerson, Joel, Sandra Harbert Petrulionis, and Laura Dassow Walls, eds. *Oxford Handbook of Transcendentalism*. Oxford: Oxford University Press, 2010.

Myerson, Joel. *The Cambridge Companion to Henry David Thoreau*. Cambridge: Cambridge University Press, 2013.

—. "Emerson's 'Thoreau': A New Edition from Manuscript." *Studies in the American Renaissance* (1979):17–92.

Newman, Lance. *Our Common Dwelling: Henry Thoreau, Transcendentalism, and the Class Politics of Nature*. New York: Palgrave Macmillan, 2005.

—. "Thoreau's Materialism: From *Walden* to *Wild Fruits*." *Nineteenth-Century Prose* 31, no. 2 (Fall 2004): 105–37.

Nicolar, Joseph. *The Life and Traditions of the Red Man*. Edited by Annette Kolodny. Durham, NC: Duke University Press, [1893] 2007.

Noble, Mark. "The Riddle of Forest Succession." In Arsic, *Dispersion*, 127–41.

Papa, James A., Jr. "Reinterpreting Myths: The Wilderness and the Indian in Thoreau's Maine Woods." *Midwest Quarterly* 40, no. 2 (Winter 1999): 215–27.

Peck, H. Daniel. *Thoreau's Morning Work: Memory and Perception in A Week on the Concord and Merrimack Rivers, the Journal, and Walden*. New Haven, CT: Yale University Press, 1990.

Petrulionis, Sandra Harbert, ed. *Thoreau in His Own Time*. Iowa City: University of Iowa Press, 2012.

Pickford, Benjamin. "*Cape Cod*, Literature, and the Illocality of Thinking about Capital." In Specq, Walls, and Nègre, *Thoreau beyond Borders*, 179–94.

Powers, Richard. *The Overstory*. New York: Norton, 2018.

Primack, Richard. *Walden Warming: Climate Change Comes to Thoreau's Woods*. Chicago: University of Chicago Press, 2014.

Quammen, David. *Song of the Dodo: Island Biogeography in an Age of Extinctions*. New York: Simon & Schuster, 1996.

Richardson, Robert D., Jr. *Henry Thoreau: A Life of the Mind*. Berkeley: University of California Press, 1986.

Robinson, David M. *Natural Life: Thoreau's Worldly Transcendentalism*. Ithaca, NY: Cornell University Press, 2004.

———. "Thoreau: Crossing to the Sacred." In Specq, Walls, and Nègre, *Thoreau beyond Borders*, 213–25.

———. "Thoreau, Modernity, and the Seasons." In Specq, Walls, and Granger, *Thoreauvian Modernities*, 59–68.

Rossi, William. "Evolution." In Finley, *Context*, 279–87.

———. "Making *Walden* and Its Sandbank," *Concord Saunterer*, 30 (2022): 10–58.

Rossi, William, ed. *Walden, Civil Disobedience, and Other Writings*. 3rd ed. New York: W. W. Norton, 2008.

Sattelmeyer, Robert. "The Remaking of *Walden*." In *Writing the American Classics*, edited by James Barbour and Tom Quirk, 53–78. Chapel Hill: University of North Carolina Press, 1990.

———. *Thoreau's Reading: A Study in Intellectual History*. Princeton, NJ: Princeton University Press, 1988.

Sayre, Robert. *Thoreau and the American Indians*. Princeton, NJ: Princeton University Press, 1977.

Shanley, James Lyndon. *The Making of Walden*. Chicago: University of Chicago Press, 1957.

Simard, Suzanne. *Finding the Mother Tree: Discovering the Wisdom of the Forest*. New York: Knopf, 2021.

Specq, François, Laura Dassow Walls, and Michel Granger, eds. *Thoreauvian Modernities: Transatlantic Conversations on an American Icon*. Athens: University of Georgia Press, 2013.

Specq, François, Laura Dassow Walls, and Julien Nègre, eds. *Thoreau beyond Borders: New International Essays on America's Most Famous Writer*. Amherst: University of Massachusetts Press, 2020.

Thoreau, Henry David. *Cape Cod*. Edited by Joseph J. Moldenhauer. Princeton, NJ: Princeton University Press, 1988.

———. *The Correspondence*. Edited by Robert N. Hudspeth, Elizabeth Hall Witherell, and Lihong Xie. *The Writings of Henry D. Thoreau*. 2 vols. Princeton, NJ: Princeton University Press, 2013–.

———. *Early Essays and Miscellanies*. Edited by Joseph J. Moldenhauer and Edwin Moser, with Alexander Kern. Princeton, NJ: Princeton University Press, 1975.

———. *Excursions*. Edited by Joseph J. Moldenhauer. Princeton, NJ: Princeton University Press, 2007.

———. *Faith in a Seed: The Dispersion of Seeds, and Other Late Natural History Writings*. Edited by Bradley P. Dean. Washington, DC: Island, 1993.

———. *The Journal of Henry David Thoreau*. 14 vols. Edited by Bradford Torrey and Francis H. Allen. Boston: Houghton Mifflin, 1906. Reprint, New York: Dover, 1962.

———. *Journal: The Writings of Henry D. Thoreau*. Edited by Elizabeth Hall Witherell, Robert Sattlemeyer, and Thomas Blanding. 8 vols. to date. Princeton, NJ: Princeton University Press, 1981–.

———. *The Maine Woods*. Edited by Joseph J. Moldenhauer. Princeton, NJ: Princeton University Press, 1972.

———. *Reform Papers*. Edited by Wendell Glick. Princeton, NJ: Princeton University Press, 1973.

———. *Translations*. Edited K. P. van Anglen. Princeton: Princeton University Press, 1986.

———. *Walden*. Edited by J. Lyndon Shanley. Princeton, NJ: Princeton University Press, 1971.

———. *Wild Fruits: Thoreau's Rediscovered Last Manuscript*. Edited by Bradley P. Dean. New York: W. W. Norton, 2000.

———. *A Week on the Concord and Merrimack Rivers*. Edited by Carl F. Hovde, William L. Howarth, and Elizabeth Hall Witherell. Princeton, NJ: Princeton University Press, 1980.

Thorson, Robert. "Physical Science." In Finley, *Context*, 247–58.

———. *The Boatman: Henry David Thoreau's River Years*. Cambridge, MA: Harvard University Press, 2017.

———. *Walden's Shore: Henry David Thoreau and Nineteenth-Century Science*. Cambridge, MA: Harvard University Press, 2014.

Traub, Courtney. "'First-Rate Fellows': Excavating Thoreau's Radical Egalitarian Reflections in a Late Draft of 'Allegash.'" *Concord Saunterer* 23 (2015): 74–96.

Vogelius, Christa Holm. "*Cape Cod*'s Transnational Bodies." In Specq, Walls, and Nègre, *Thoreau beyond Borders*, 25–39.

Walls, Laura Dassow. "A Material Faith: Thoreau's Terrennial Turn." In Arsic, *Dispersion*, 37–58.

———. "Counter Frictions: Thoreau and the Integral Commons." In Specq, Walls, and Nègre, *Thoreau beyond Borders*, 226–42.

———. *Material Faith: Thoreau on Science*. Boston: Houghton Mifflin, 1999.

———. *Seeing New Worlds: Henry David Thoreau and Nineteenth-Century Natural Science.* Madison: University of Wisconsin Press, 1995.

———. "Technology." In Finley, *Context,* 165–74.

———. *The Daily Henry David Thoreau: A Year of Quotes from the Man Who Lived in Season.* Chicago: University of Chicago Press, 2020.

———. *The Passage to Cosmos: Alexander von Humboldt and the Shaping of America.* Chicago: University of Chicago Press, 2009.

———. *Thoreau: A Life.* Chicago: University of Chicago Press, 2017.

———. "Walking West, Gazing East: Planetarity on the Shores of Cape Cod." In Specq, Walls, and Granger, *Thoreauvian Modernities,* 25–40.

Warren, James Perrin. *Culture of Eloquence: Oratory and Reform in Antebellum America.* State College: Penn State University Press, 1999.

Weisburg, Michael C. "Beyond the Borders of Time: Thoreau and the 'Ante-Pilgrim' History of the New World." In Specq, Walls, and Nègre, *Thoreau beyond Borders,* 40–54.

West, Michael. "Charles Kraitsir's Influence upon Thoreau's Theory of Language." *ESQ* 19 (1973): 262–74.

Wohlleben, Peter. *The Hidden Life of Trees: What They Feel, How They Communicate— Discoveries from a Secret World.* Vancouver: Greystone, 2015.

Wood, Alphonso. *A Class-Book of Botany.* 2nd ed. Boston: Crocker & Brewster, 1848.

Index of Plants

Index of Subjects

HDT refers to Henry David Thoreau.

Adam Trexler • *Anthropocene Fictions: The Novel in a Time of Climate Change*

Kate Rigby • *Dancing with Disaster: Environmental Histories, Narratives, and Ethics for Perilous Times*

Byron Caminero-Santangelo • *Different Shades of Green: African Literature, Environmental Justice, and Political Ecology*

Jennifer K. Ladino • *Reclaiming Nostalgia: Longing for Nature in American Literature*

Dan Brayton • *Shakespeare's Ocean: An Ecocritical Exploration*

Scott Hess • *William Wordsworth and the Ecology of Authorship: The Roots of Environmentalism in Nineteenth-Century Culture*

Axel Goodbody and Kate Rigby, editors • *Ecocritical Theory: New European Approaches*

Deborah Bird Rose • *Wild Dog Dreaming: Love and Extinction*

Paula Willoquet-Maricondi, editor • *Framing the World: Explorations in Ecocriticism and Film*

Bonnie Roos and Alex Hunt, editors • *Postcolonial Green: Environmental Politics and World Narratives*

Rinda West • *Out of the Shadow: Ecopsychology, Story, and Encounters with the Land*

Mary Ellen Bellanca • *Daybooks of Discovery: Nature Diaries in Britain, 1770–1870*

John Elder • *Pilgrimage to Vallombrosa: From Vermont to Italy in the Footsteps of George Perkins Marsh*

Alan Williamson • *Westernness: A Meditation*

Kate Rigby • *Topographies of the Sacred: The Poetics of Place in European Romanticism*

Mark Allister, editor • *Eco-Man: New Perspectives on Masculinity and Nature*

Heike Schaefer • *Mary Austin's Regionalism: Reflections on Gender, Genre, and Geography*

Scott Herring • *Lines on the Land: Writers, Art, and the National Parks*

Glen A. Love • *Practical Ecocriticism: Literature, Biology, and the Environment*

Ian Marshall • *Peak Experiences: Walking Meditations on Literature, Nature, and Need*

Printed in the USA
CPSIA information can be obtained
at www.ICGtesting.com
LVHW090059030923
756292LV00024B/198

9 780813 949482